Springer Theses

Recognizing Outstanding Ph.D. Research

Aims and Scope

The series "Springer Theses" brings together a selection of the very best Ph.D. theses from around the world and across the physical sciences. Nominated and endorsed by two recognized specialists, each published volume has been selected for its scientific excellence and the high impact of its contents for the pertinent field of research. For greater accessibility to non-specialists, the published versions include an extended introduction, as well as a foreword by the student's supervisor explaining the special relevance of the work for the field. As a whole, the series will provide a valuable resource both for newcomers to the research fields described, and for other scientists seeking detailed background information on special questions. Finally, it provides an accredited documentation of the valuable contributions made by today's younger generation of scientists.

Theses are accepted into the series by invited nomination only and must fulfill all of the following criteria

- They must be written in good English.
- The topic should fall within the confines of Chemistry, Physics, Earth Sciences, Engineering and related interdisciplinary fields such as Materials, Nanoscience, Chemical Engineering, Complex Systems and Biophysics.
- The work reported in the thesis must represent a significant scientific advance.
- If the thesis includes previously published material, permission to reproduce this must be gained from the respective copyright holder.
- They must have been examined and passed during the 12 months prior to nomination.
- Each thesis should include a foreword by the supervisor outlining the significance of its content.
- The theses should have a clearly defined structure including an introduction accessible to scientists not expert in that particular field.

More information about this series at http://www.springer.com/series/8790

Giulia Cozzani

Microphysics of Magnetic Reconnection in Near-Earth Space

Spacecraft Observations and Numerical Simulations

Doctoral Thesis accepted by
Laboratoire de Physique des Plasmas—École Polytechnique, Palaiseau, France

Springer

Author
Dr. Giulia Cozzani
Swedish Institute of Space Physics
Uppsala, Sweden

Supervisors
Dr. Alessandro Retinò
Laboratoire de Physique des
Plasmas—École Polytechnique
Palaiseau, France

Prof. Francesco Califano
Dipartimento di Fisica 'E.Fermi'
Università di Pisa
Pisa, Italy

Dr. Patrick Canu
Laboratoire de Physique des
Plasmas—École Polytechnique
Palaiseau, France

ISSN 2190-5053 ISSN 2190-5061 (electronic)
Springer Theses
ISBN 978-3-030-56141-3 ISBN 978-3-030-56142-0 (eBook)
https://doi.org/10.1007/978-3-030-56142-0

This Springer imprint is published by the registered company Springer Nature Switzerland AG
The registered company address is: Gewerbestrasse 11, 6330 Cham, Switzerland

Tanto più è grande quello che si cerca
tanto più nel piccolo esso si mostra
(found on the door lintel of my rented room in Pisa)

To my parents, Gianna & Ettore

Supervisor's Foreword

Magnetic reconnection is a fundamental process occurring in plasmas during which changes in the topology of magnetic fields lead to the conversion of magnetic energy into energy of charged particles, resulting in major particle acceleration. Reconnection is ubiquitous: it is observed in laboratory and space plasmas and it is thought to play a major role in distant astrophysical plasmas. Examples range from tokamaks and laboratory plasma devices on Earth, planetary magnetospheres, the interplanetary space, solar and stellar coronae, the interstellar medium, neutron stars, accretion disks, astrophysical jets and galaxy clusters.

Despite of the fact that reconnection has been studied for many decades, both theoretically and experimentally, a number of outstanding science questions are still open. Important examples are: the physics of reconnection at electron scales (where reconnection is initiated), the relationship between reconnection and turbulence and the role of reconnection for particle acceleration. These problems are among the most studied topics of current plasma physics research.

Studying and understanding reconnection from an experimental point of view require direct observations of electromagnetic fields and ion and electron distribution functions. The near-Earth space is the best natural laboratory where such observations can be obtained, through high-resolution in situ spacecraft measurements; in particular, multi-spacecraft measurements which allow to study reconnection at different plasma scales. Due to the complexity of reconnection physics, numerical simulations are of pivotal importance to interpret in situ measurements, especially kinetic simulations which are able to fully resolve the complex dynamics of ion and electrons. The combination of in situ measurement and kinetic simulation is the most powerful tool to study reconnection physics and address its outstanding open questions.

This thesis presents recent advances on the physics of magnetic reconnection, namely on the physics at electron scales, which is investigated through both in situ spacecraft observations in the Earth's magnetosphere and novel kinetic numerical simulations. The thesis focuses on the physics of the so-called Electron Diffusion Region (EDR), which is the microscopic region where reconnection is initiated and where electrons are decoupled from the magnetic field and are accelerated.

In a first study, the thesis takes advantage of the very recent NASA's Magnetospheric MultiScale Mission (MMS), launched in 2015, which have allowed to obtain for the first time in situ measurements at electron scales. In this study, the structure of the EDR at the Earth's magnetopause is investigated in order to understand the structure of this region at electron scales, which is still not fully understood. By using the unique capability of MMS to provide simultaneous measurements at four points separated by electron-scale distances, this study is able to provide novel evidence for an inhomogeneous structure of the EDR, which appears to be much more complex than the laminar structure suggested by previous in situ observations and simulations. Energy conversion in the EDR is found to be patchy, with regions where the magnetic energy is converted into electron heating as expected, but also regions where the opposite occurs and strong electromagnetic fluctuations are generated. These results suggest that, at least in some cases, the structure of the EDR may be turbulent, possibly resulting in a more efficient energy conversion and electron acceleration.

In situ MMS measurements provide a unique tool to study the structure of the EDR. Yet, due to the complexity of the plasma physics in this region, kinetic simulations are crucial to support data interpretation. In particular, the observational results from the first study strongly indicate the need of kinetic simulations with very high spatial resolution and very low noise, in order to resolve the complex structure of the EDR and of electron dynamics therein. Driven from the observational results, a second study has been performed to address the physics of the EDR through a novel, fully-kinetic Eulerian Vlasov-Darwin numerical code (ViDA) which has been developed during the thesis. The ViDA code is able to resolve with very high resolution the phase-space details of the electron distribution functions, allowing to make unprecedented study of electron dynamics. The thesis includes a full explanation of the ViDA code and its benchmark tests. The code is later used to study the contribution of the different terms in the generalized Ohm's law within the EDR, highlighting the role of the electron inertia term for decoupling the electrons from the magnetic field.

The combination of very recent MMS in situ measurements with the novel, fully-kinetic ViDA code used in this thesis has allowed to obtain remarkable novel results on the physics of reconnection at electron scales, namely on the structure of the Electron Diffusion Region. These results have important implications also to understand the role of turbulence in reconnection as well as the mechanisms of electron acceleration during reconnection. Overall, the thesis makes an important contribution to advance the understanding of reconnection physics in space plasmas, with far reaching implication for laboratory and astrophysical plasmas too.

Palaiseau, France Dr. Alessandro Retinò
June 2020

Preface

Magnetic reconnection is a fundamental energy conversion process occurring in space and laboratory plasmas. Reconnection takes place in thin current sheets leading to the reconfiguration of magnetic field topology and to conversion of magnetic energy into acceleration and heating of particles. Today reconnection is recognized to play a key role in the Earth-solar environment, from the solar corona to the solar wind, to magnetosheath, at the Earth's magnetopause, and in the magnetotail. Reconnection is initiated in a microscopic region, called the Electron Diffusion Region (EDR), where electrons decouple from the magnetic field and are energized by electric fields. Despite the very significant advances that have been made in the understanding of the magnetic reconnection process by means of in-situ measurements (notably provided by the Cluster spacecraft mission) and by numerical simulations, the small electron scale physics of the dissipation region remains basically unsolved. It is only the last years, with the launch of the Magnetospheric MultiScale mission (MMS) together with the recent impressive increasing of computational capabilities of supercomputers, that the dynamics of the Electron Diffusion Region has started to be enlightened. One of the key, yet still open questions, is whether the Electron Diffusion Region has a preferred homogeneous or inhomogeneous structure at electron scales and below. The purpose of this Thesis is to advance in the understanding of the structure of the Electron Diffusion Region using two different approaches, notably MMS spacecraft observations and kinetic full Vlasov simulations.

The material presented in the Thesis has been organized in order to provide the fundamental theoretical concepts and to present the state of the art about Electron Diffusion Region observations and fully-kinetic Vlasov codes before discussing the results. In particular, after the Introduction (Chap. 1), Chapter 2 provides a description of the magnetic reconnection process at large scale, adopting the traditional Magneto-Hydro-Dynamics (MHD) theoretical framework. Despite the fact that the MHD model overlooks the kinetic effects, it still provides important results especially when dealing with global and large-scale aspects of the process. Then, an overview of reconnection at the Earth's magnetosphere is provided in Chap. 3 since the physical context where magnetic reconnection is investigated in this Thesis is the Earth's magnetosphere and the magnetopause in particular.

Before discussing the properties of the Electron Diffusion Region, two Chapters (Chaps. 4 and 5) are devoted to the discussion of the methodology for both spacecraft data and numerical simulations. In particular, Chap. 4 describes the Magnetospheric MultiScale (MMS) mission providing information about the spacecraft, the instrumentation on board, the orbit and data products and it presents the data analysis techniques that are used for the data analysis reported in Chap. 7. Chapter 5 gives a rapid overview of the numerical models which are commonly used to study magnetic reconnection focusing in particular on the state of the art of full kinetic Vlasov simulations. Chapter 6 summarizes the main observational and numerical studies about the Electron Diffusion Region and it focuses in particular on the operational criteria of identification in the context of spacecraft data.

The following Chapters present the results. Chapter 7 presents MMS observations of an Electron Diffusion Region encounter at the subsolar magnetopause when the four MMS probes were located at the smallest interspacecraft separation of $\sim$6 km, which is comparable to a few electron inertial length ($d_e \sim 2$ km). We find that the Electron Diffusion Region is rather inhomogeneous and that the pattern of the energy conversion is patchy, showing that the structure of the Electron Diffusion Region at the magnetopause can be much more complex than it has been found in other Electron Diffusion Region encounters reported by MMS and more complex than it is usually depicted by kinetic Particle-In-Cell (PIC) simulations. Our MMS data analysis has pointed out the need of simulations with better spatial resolution and low noise on the electron scales, in particular on the electric field, in order to better understand the kinetic physics at play. Following this motivation, the following part of the Thesis aims at studying the Electron Diffusion Region by using a novel fully-kinetic Eulerian Vlasov-Darwin model which we have implemented in the numerical ViDA code. Such code, presented in Chap. 8, allows to perform low noise and high-resolution simulations and it is specifically designed to improve our understanding of the kinetic dynamics of collisionless plasma at electron scales by giving access to the fine phase space details of the electron distribution function. A first part is devoted to the testing of the code by performing 2D symmetric magnetic reconnection simulations (Chap. 8). The next step has been to use the ViDA code to investigate another long-standing topic in the context of reconnection, namely the contribution of the different terms of Ohm's law in sustaining the reconnection electric field in the Electron Diffusion Region. In particular, the analysis focuses on the contribution of the electron inertia term which is responsible for the decoupling of the electron dynamics from the magnetic field. Results are shown in Chap. 9. Conclusions are provided in Chap. 10 together with perspectives for possible future observational and numerical work.

Uppsala, Sweden Giulia Cozzani

List of Publications

This thesis is based on the following papers

- G. Cozzani *et al.*, *In situ spacecraft observations of a structured electron diffusion region during magnetopause reconnection*, Phys. Rev. E **99**, 043204 (2019).
- O. Pezzi, G. Cozzani *et al.*, *ViDA: a Vlasov-DArwin solver for plasma physics at electron scales*, J. Plasma Phys. **85**, 905850506 (2019).

The following papers are not included in this Thesis

- H. Breuillard *et al.*, *The properties of lion roars and electron dynamics in mirror mode waves observed by the Magnetospheric MultiScale mission*, J. Geophys. Res. **123**, 93–103 (2018).
- J. M. Webster *et al.*, *Magnetospheric Multiscale Dayside Reconnection Electron Diffusion Region Events*, J. Geophys. Res. **123**, 4858–4878 (2018).
- Z. Wang *et al.*, *Electron distribution functions around a reconnection X-line resolved by the FOTE method, Geophys.* Res. Lett. **46**, 1195–1204 (2019).
- Z. Z. Chen *et al.*, *Electron-driven dissipation in a tailward flow burst*, Geophys. Res. Lett. **46**, 5698–5706 (2019).

Acknowledgements

A *cotutelle* Ph.D. project doubles the troubles but it also doubles the opportunities to meet and work with people of great value.

First and foremost, thank you to my three advisors Alessandro Retinò, Francesco Califano and Patrick Canu. I'm deeply grateful for your patience and your guidance. Alessandro and Francesco, thank you for believing in this project that has been ambitious, challenging and exciting for me. It has been a great opportunity to work with people that have such a competence and experience. You gave me the possibility to grow in all sense and to explore. Thank you Patrick for your support and your time.

I would like to express my gratitude to Karine Bocchialini, Roberto Bruno, Stefano Markidis and Hantao Ji, who agreed to be members of my Ph.D. committee. It has been a real honor for me. In particular, I would like to thank the two referees, Stefano Markidis and Hantao Ji, for the interest they showed in my research and for having read my manuscript so thoroughly.

I would like to thank all the people of the *Plasmas Spatiaux* group of the Laboratoire de Physique des Plasmas (LPP). In particular, I would like to thank Olivier Le Contel—who has always been always ready to discuss and listen—Hugo Breuillard, Dominique Fontaine, Laurent Mirioni and many others. I am grateful for the opportunity of the conferences participation provided by LPP. Thank you to the *équipe informatique* and the *gestionnaires* of the lab for their availability and kindness. I'm grateful to Anne Bourdon and Roch Smets who have been the members of my *comité de suivi de thèse*.

Thank you to Alexandra Alexandrova for all the support, the enthusiasm, the great discussions about science (and art and music and life and ...) and all the positive energy. Filomena and Sae, it has been a pleasure to share the office with you. Thank you for all the good discussions and the good moments in Paris and in other locations, notably in Cosenza for Filomena's defense.

I would like to thank the *plasmisti* group of Pisa: Sid Fadanelli, Paula S. Kleij, Francesco *Paco* Finelli, Manuela Sisti, Jérémy Dargent, Roberto Manuzzo, Giuseppe *Peppe* Arrò, Federico *Job* Lavorenti, Louis Richard, Francesco Pisani. Thank you for all the interesting conversations about science as well as about everything else, for the *panuozzi* and the nice toasts to our achievements accompanied by good food (imposed as a group rule by Francesco Califano). Thank you to Francesco Pegoraro and Fulvio Cornolti for their availability to discuss. A special acknowledgment goes to Pasticceria Frangioni since without its amazing coffee and cakes a Ph.D. student's life could become pretty much unbearable.

Thank you to Oreste Pezzi. Your contribution has been critical and a large part of what I have learned about simulations came from you. Thank you for your patience and your enthusiasm, working on ViDA has been really enriching for me.

Thank you to Yuri Khotyaintsev and Andris Vaivads who kept asking difficult questions that helped me improve my understanding of the spacecraft data.

I'm grateful to Huishan Fu that gave me the unique opportunity to visit Beihang University in Beijing for a month. Thank you to the Ph.D. and master students in Beihang, who welcomed me in the nicest way.

This thesis has been possible also because of the hard work of the members of the MMS team. Thank you for all your efforts to provide high quality data and for creating a great scientific community.

I am grateful to the staff of CINECA and in particular to Carlo Cavazzoni and Massimiliano Guarrasi that have always been ready to help me when using MARCONI.

A Ph.D. in joint supervision in two different countries brings you to say "goodbye" and "I'll miss you" too often. However, in this years I have learnt that friendship does not fade away that easily when it is deep-rooted, as in France as in Italy.

I'm grateful to Bianca Pierattini, Chiara Colli, Giulia Borrini, Irene Carabelli, Valentina Cattafesta, Greta Maggiani that have been there for many years now. Thank you for everything. Thank you to Paolo Erdman and Francesco Macheda for the scientific, political and philosophical discussions which always have a high level of *polemica*.

Thank you to the *Deutsch leftovers* (Erica Telford, Sofia Pombo, Christiane Brenner, Mariana Ponce de León, Andrea Napoli (M.)) and the other people from the 1st *floor Pasteur* with whom I shared happy moments in Paris and who could totally relate with the struggles of the Ph.D. life.

I'm deeply grateful to my family that has always supported me and believed in me. In particular, I would like to thank my grandmother Carmela (Carmen) Merini and my uncle Fabio Codeglia, who are an example of strength and honesty for me.

My thoughts are with my parents, Gianna Codeglia and Ettore Cozzani. Simply enough, I wouldn't be who I am today if it wasn't for you. I am deeply grateful to you.

At last, I would like to thank Gian Marcello Andolina. I could not imagine this journey without you. Thank you Marci, there is still such a lot of world to see together.

Contents

Useful Symbols and Formulas

Symbols

d_e	electron inertial length, [km]
$d_i(d_p)$	ion (proton) inertial length, [km]
ρ_e	electron gyroradius, [km]
$\rho_i(\rho_p)$	ion (proton) gyroradius, [km]
λ_D	Debye length, [km]
$v_{th,e}$	electron thermal speed, [km/s]
$v_{th,i}(v_{th,p})$	ion (proton) thermal speed, [km/s]
$V_{A,e}$	electron Alfvén speed, [km/s]
$V_{A,i}(V_{A,p})$	ion (proton) Alfvén speed, [km/s]
$\omega_{p,e}$	electron plasma frequency, [s^{-1}]
$\omega_{p,i}(\omega_{p,p})$	ion (proton) plasma frequency, [s^{-1}]
$\Omega_{c,e}$	electron cyclotron frequency, [s^{-1}]
$\Omega_{c,i}(\Omega_{c,p})$	ion (proton) cyclotron frequency, [s^{-1}]
$T_e(T_{e,\|\|}, T_{e,\perp})$	electron temperature, [eV] (parallel and perpendicular to the magnetic field)
$T_i(T_{i,\|\|}, T_{i,\perp})$	ion (proton)temperature, [eV] (parallel and perpendicular to the magnetic field)
η	resistivity, [Ω m]
β	ratio of the plasma pressure to the magnetic pressure
$\sqrt{Q}$	electron pressure agyrotropy
R_m	magnetic Reynolds number
c	light speed, 3×10^5 [km/s]
e	electron charge, 1.6×10^{-19} [C]
m_e	electron mass, 9.1×10^{-31} [kg]
$m_i(m_p)$	ion (proton) mass, 1.6×10^{-27} [kg]
R_E	equatorial radius of Earth, 6371 [km]

Formulas

$\omega_{p,\alpha} = (4\pi n_\alpha e^2/m_\alpha)^{1/2} \quad d_\alpha = c/\omega_{p,\alpha},\ \alpha = e, i, p$

$\lambda_D = v_{th,e}/\omega_{p,e}$

$\Omega_{c,\alpha} = eB/m_\alpha c \quad \rho_\alpha = v_{th,\alpha}/\Omega_{c,\alpha}$

$V_{A,\alpha} = B/(4\pi n_\alpha m_\alpha)^{1/2}$

List of Figures

List of Tables

Chapter 1
Introduction

A plasma is defined as a gas composed of charged particles, globally neutral and dominated by electromagnetic forces.

Plasma state is usually referred to as the fourth state of matter since a plasma can be thought to be produced by heating an ordinary gas to such temperature that the amount of charged particles drastically affects the system dynamics. Being composed by charged particles, a plasma is characterized by two competing tendencies that neutral gases do not display. The action of the electromagnetic forces leads to a tendency to cohesion while the chaotic movement of the particles provokes a tendency to disaggregate. The interaction of charged particles affects the particles motion, providing the coherent and collective behavior of the plasma.

Studying a plasma system means dealing with a self-consistent problem: the particles of the plasma react collectively to forces exerted by electric and magnetic fields; the fields are modified by the charge densities and current densities of the plasma. The dynamics of the system is modeled, in general, using a kinetic description based on the Boltzmann equation coupled to Maxwell equations. However, depending on the typical wavelength and time scale of the specific process that has to be analyzed, different approaches can be adopted. The fluid or Magneto-Hydro-Dynamics (MHD) approach has proved valuable in studying the global aspects of the system. Of course, such an approach based on a macroscopic view of the plasma ceases to be valid as soon as the system develops fluctuations comparable to the ion gyroradius.

Plasma may not be the dominant state of matter that is found in nature on Earth, but more than 99% of all known matter in the Universe is in the plasma state (Baumjohann and Treumann 1996). For this reason, it is crucial to study astrophysical and space plasmas, which include plasma in the Earth's magnetosphere and the solar wind. These plasmas are usually highly magnetized and collisionless so that the kinetic description is based on the Vlasov equation coupled to the Maxwell equations.

G. Cozzani, *Microphysics of Magnetic Reconnection in Near-Earth Space*, Springer Theses, https://doi.org/10.1007/978-3-030-56142-0_1

One of the fundamental processes that is ubiquitous in plasmas is magnetic reconnection, a major cause of energy conversion and transport in plasma. Reconnection is recognized to play a key role in the Earth-solar environment – from the solar wind (Phan et al. 2006), to magnetosheath (Phan et al. 2018; Retinò et al. 2007), at the Earth's magnetopause (Burch et al. 2016a; Mozer et al. 2002), and in the magnetotail (Øieroset et al. 2001; Torbert et al. 2018) – but also in the solar corona (Cargill 2015), in magnetized accretion disks around black holes (de Gouveia Dal Pino and Lazarian 2000) and in magnetars (Lyutikov 2003).

Magnetic reconnection was firstly suggested by R. Giovanelli as a mechanism that could explain the intense energy release observed during solar flares. Giovanelli realized that solar flares are associated with sunspots (Fig. 1.1, Giovanelli 1939) and that the magnetic field near a sunspot is characterized by a "neutral point" where the magnetic field goes to zero. The presence of the neutral point is suggested as the necessary condition for allowing the particles of the plasma to be accelerated by the electric field and to reach the high energies observed during solar flares (Giovanelli 1947).

This idea was applied also to the magnetosphere providing a potential mechanism to explain the accelerated particles giving rise to aurora (Hoyle 1950). In this context, J. Dungey described the topological rearrangement of field lines that takes place when a neutral point is present and that causes particle acceleration (Dungey 1953). This concept was then applied specifically to the magnetosphere and the outcome was the conception of the Dungey cycle (Dungey 1961) (see Sect. 3.1).

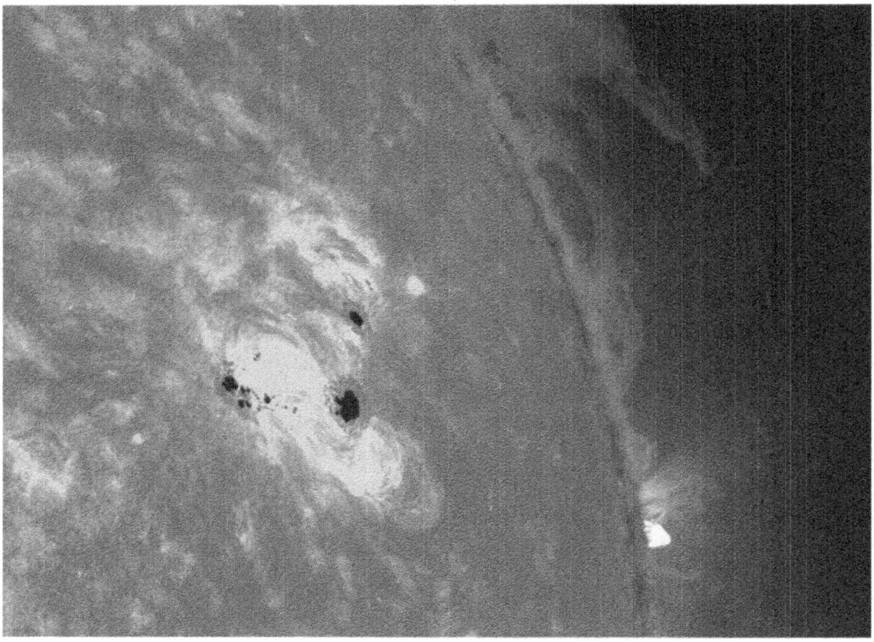

Fig. 1.1 Active region on the Sun. The sunspots are the darker areas. Image credit: NASA/SDO/AIA/HMI/Goddard Space Flight Center. https://svs.gsfc.nasa.gov/11211

The first model aiming to explain the origin of this process was proposed by Sweet (1958). He suggested the Ohmic dissipation to be responsible for reconnection of current sheets. Parker (1957) formulated the model of magnetic reconnection using the scaling relations in the pre-reconnecting and after-reconnecting regions. However, it was soon realized that the reconnection rate obtained with this model was too small compared to solar flares observations (Parker 1963). Another model was developed by Petschek (1964) who proposed that besides the mechanism of magnetic field diffusion in the reconnection region, the energy can be converted at shock waves. In the framework of Petschek model, the finite conductivity is assumed to be localized within a diffusion region which is much smaller that the whole current sheet and that leads to the formation of two pairs of shock waves separating the inflow and outflow regions. The reconnection process in this model has a rate which is faster and closer to observations (see also (Vasyliunas 1975), a review unifying the reconnection models). However, the underlying kinetic mechanisms which lead to the enhanced localized resistivity featured in Petschek model are still unknown and the localized resistivity has not yet been verified experimentally.

In 1963, reconnection was proven to play a role also in fusion devices (Furth et al. 1963). Furth et al. (1963) showed by analytical calculations that reconnection can operate via the tearing instability which in turn leads the current sheet to break into a series of magnetic islands or plasmoids (Ji and Daughton 2011; Priest and Forbes 2000).

The first observational evidences of reconnection operating at the Earth's magnetosphere were *indirect* evidences. For instance, it was shown that the interplanetary magnetic field direction is related to auroral and geomagnetic activity (Fairfield and Cahill Jr 1966); that a southward interplanetary magnetic field causes the magnetopause to shift inward (the so-called magnetospheric erosion (Aubry et al. 1970)); that energetic particles from the Sun reach the magnetic field lines anchored at the polar cap first (Fennell 1973).

Then, the first strong direct proofs for reconnection were produced by means of observations of the ISEE spacecraft (Paschmann et al. 1979; Sonnerup et al. 1981).

It has become clear that the terrestrial magnetosphere is the perfect natural laboratory to study the magnetic reconnection process. Indeed, after the first observations provided by the ISEE spacecraft, very significant advances have been made in the detailed understanding of the magnetic reconnection process by means of high quality in situ measurements, notably with data from Polar, Cluster, THEMIS/ARTEMIS and TWINS (Yamada et al. 2010). Finally, the launch of the Magnetospheric Multi-Scale (MMS) mission has given a new impulse to the investigation of the magnetic reconnection process by providing particles data with the adequate high resolution to analyze the electron scales (Burch et al. 2016b).

In addition, the Earth's magnetosphere is an environment with an extremely large variety of conditions in terms of density, temperature, value of the β parameter (which corresponds to the ratio between the thermal and the magnetic pressure). For this reason, the progress made in the understanding magnetic reconnection at the Earth's magnetosphere can be exported to investigate magnetic reconnection occurring in other environments (Vaivads et al. 2009) (see Fig. 1.2).

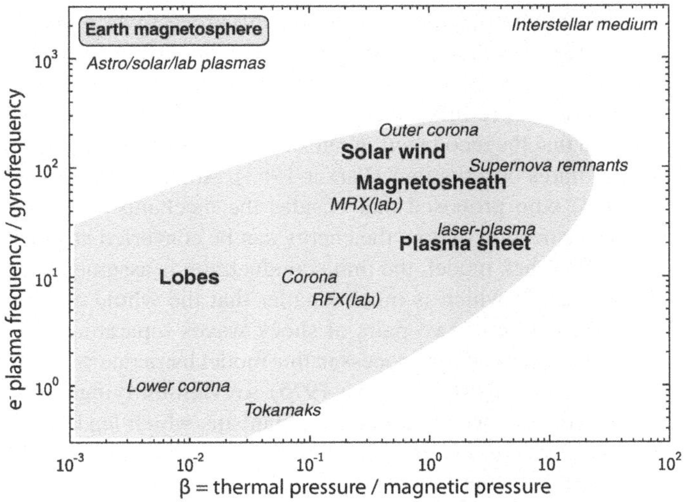

Fig. 1.2 Different kind of plasmas are located in a parameter space composed by the plasma β and the ratio between the electron plasma frequency $\omega_{p,e}$ and the electron cyclotron frequency $\Omega_{c,e}$. β mainly describes ion physics while $\omega_{p,e}/\Omega_{c,e}$ characterizes electron processes. Once plotted in a dimensionless parameters space, different plasmas appear to be close to each other. Also, the parameter range that is covered by the conditions that are found at the magnetosphere (colored area) is especially large. Adapted from (Vaivads et al. 2009). The figure is used with permission from A. Vaivads

Magnetic reconnection has been also largely studied by means of laboratory experiments, notably the Magnetic Reconnection Experiment (MRX) (Ji et al. 2008; Yamada et al. 1997). Laboratory experiments have the virtue to be repeatable and to have controlled boundary conditions while in situ observations have to rely on *encounters* of the sought process which are to some extent fortuitous. However, investigating magnetic reconnection with in situ measurements has some advantages which include the fact that there are no artificial boundaries, the electrons and ions distribution functions can be measured directly and the spatial resolution of the fields measurements can be below the Debye length (Vaivads et al. 2009).

Understanding magnetic reconnection is crucial because it is far to be a topic of interest only in the field of space plasmas. In particular, magnetic reconnection is found to be a limiting factor in achieving and maintaining electron temperatures high enough to initiate fusion in magnetic-confinement fusion devices (Yamada et al. 1994). Furthermore, magnetic reconnection plays a substantial role also in the context of the space weather, eventually affecting the interior of the magnetosphere and even the surface of the Earth. The magnetosphere acts as a protective shield against the energetic particles of solar flares and cosmic rays. However, this barrier is not impenetrable. Reconnection at the dayside magnetosphere is responsible for the coupling of the solar sourced plasma of the solar wind and the magnetospheric plasma allowing the penetration of energetic particles in the magnetospheric environment.

This coupling is especially strong when the Sun is active, producing flares associated with Coronal Mass Ejections (CME) that interact with the terrestrial magnetic environment. The effects can be registered in space (damaged spacecraft, communication failures, radiation risk for astronauts), in the atmosphere (radio communication problems, diversion of transpolar flights) and even on ground (interference with power grids).

Despite the significant advances that have been done in about 80 years of research from the first ideas of Giovanelli (1939), some fundamental aspects of magnetic reconnection in collisionless plasmas are still poorly understood. As it will be widely discussed in the following, magnetic reconnection is a multiscale process which involves several characteristic lengths and time scales. Its multiscale nature makes it difficult to describe the process analytically but also to investigate it with computational and observational approaches. The topological rearrangement and the consequent energy release – which are the main characteristics of reconnection – are thought to take place in the so-called electron diffusion region which has a characteristic length comparable to the electron inertial length. The electron scale physics of the process has become accessible only recently both in numerical simulations and observations.

The main goal of my Thesis, in its broadest sense, is to advance in the understanding of the electron scale process in the diffusion region of magnetic reconnection. In order to do that, two different approaches will be used, namely the analysis of high resolution spacecraft data from the MMS mission at the Earth's magnetopause and low-noise simulation data issued from a new code based on a fully-kinetic Eulerian Vlasov-Darwin model.

As discussed in Chaps. 6 and 7, several numerical and observational studies has been focused on the electron diffusion region. However, it is not fully understood whether the EDR has a preferred homogeneous or inhomogeneous structure at electron scales and below in terms of current densities and electric fields. Throughout this Thesis, I will try to answer to this long standing question. A non laminar structure would affect the energy conversion in the diffusion region and it could influence the magnetic reconnection process globally (e.g. affecting the reconnection rate).

Concerning the units used in this Thesis, I would like to warn the reader that I have maintained the unit system that is typically used for each specific approach. In particular, MMS data are provided in MKS units while models used in numerical codes are written using the cgs system. Readers are assumed to be already familiar with this situation when reading papers about numerical simulations and spacecraft data.

Throughout this Thesis the words *ion* and *proton* will be extensively used. In the context of my research, there is no real distinction between these two terms since the plasma that are taken into account in this Thesis are composed by protons and electrons. However, while this is strictly true for simulation data since I set the atomic number of ions to 1 in the equations of the model, the Fast Plasma Investigation (FPI) particle instrument on board of MMS measures the distribution functions of electrons and ions but it does not provide information about the composition. Since ion flows into the magnetosphere can often be assumed to have solar wind-like composition of ~96% protons and >3.9% alpha particles we will use the terms *ion* and *proton* indistinctly.

References

M.P. Aubry, C.T. Russell, M.G. Kivelson, J. Geophys. Res. **75**, 7018 (1970)
W. Baumjohann, R.A. Treumann, *Basic Space Plasma Physics* (Imperial College Press, London, 1996)
J.L. Burch et al., Science **352**, aaf2939 (2016)
J.L. Burch, T.E. Moore, R.B. Torbert, B.L. Giles, Space Sci. Rev. **199**, 5 (2016)
P. Cargill, in *Magnetic Reconnection in the Solar Corona: Historical Perspective and Modern Thinking*, ed. by S. Cowley FRS, D. Southwood, S. Mitton. Magnetospheric Plasma Physics: The Impact of Jim Dungey's Research. Astrophysics and Space Science Proceedings (Springer, Cham, 2015), vol. 41
E.M. de Gouveia Dal Pino, A. Lazarian, Astrophys. J. **536**, L31 (2000)
J.W. Dungey, London, Edinburgh Dublin Philos. Mag. J. Sci. **44**, 725 (1953)
J.W. Dungey, Phys. Rev. Lett. **6**, 47 (1961)
D.H. Fairfield, L.J. Cahill Jr., J. Geophys. Res. **71**, 155 (1966)
J.F. Fennell, J. Geophys. Res. **78**, 1036 (1973)
H.P. Furth, J. Killeen, M.N. Rosenbluth, Phys. Fluids **6**, 459 (1963)
R.G. Giovanelli, Astrophys. J. **89**, 555 (1939)
R.G. Giovanelli, Mon. Not. R. Astron. Soc. **107**, 338 (1947)
F. Hoyle, Q.J.R. Meteorol, Soc. **76**, 112 (1950)
H. Ji, W. Daughton, Phys. Plasmas **18**, 111207 (2011)
H. Ji, Y. Ren, M. Yamada, S. Dorfman, W. Daughton, S.P. Gerhardt, Geophys. Res. Lett. **35**, L13106 (2008)
M. Lyutikov, Mon. Not. R. Astron. Soc. **346**, 540 (2003)
F.S. Mozer, S.D. Bale, T.D. Phan, Phys. Rev. Lett. **89**, 015002 (2002)
M. Øieroset, T.D. Phan, M. Fujimoto, R.P. Lin, R.P. Lepping, Nature (London) **412**, 414 (2001)
E.N. Parker, J. Geophys. Res. **62**, 509 (1957)
E.N. Parker, Astrophys. J. Suppl. Ser. **8**, 177 (1963)
G. Paschmann et al., Nature (London) **282**, 243 (1979)
H.E. Petschek, in *Magnetic Field Annihilation*, ed. by W.N. Hess. The Physics of Solar Flares Proceedings of the AAS-NASA Symposium (NASA Spec. Publ., 1964), vol. 50, p. 425
T.D. Phan et al., Nature (London) **439**, 175 (2006)
T.D. Phan et al., Nature (London) **557**, 202 (2018)
E.R. Priest, T. Forbes, *Magnetic reconnection. MHD theory and applications* (Cambridge University Press, Cambridge, 2000)
A. Retinò, D. Sundkvist, A. Vaivads, F.S. Mozer, M. André, C.J. Owen, Nat. Phys. **3**, 235 (2007)
B.U.Ö. Sonnerup et al., J. Geophys. Res. **86**, 10049 (1981)
P.A. Sweet, in *The Neutral Point Theory of Solar Flares*, ed. by B. Lehnert. Electromagnetic Phenomena in Cosmical Physics Proceedings from IAU Symposium no. 6 (Cambridge University Press, Cambridge, 1958), p. 123
R.B. Torbert et al., Science **362**, 1391 (2018)
A. Vaivads, A. Retinò, M. André, Plasma Phys. Control. Fusion **51**, 124016 (2009)
V.M. Vasyliunas, Rev. Geophys. **13**, 303 (1975)
M. Yamada, F.M. Levinton, N. Pomphrey, R. Budny, J. Manickam, Y. Nagayama, Phys. Plasmas **1**, 3269 (1994)
M. Yamada et al., Phys. Plasmas **4**, 1936 (1997)
M. Yamada, R. Kulsrud, H. Ji, Rev. Mod. Phys. **82**, 603 (2010)

Chapter 2
Fundamental Concepts Associated with Magnetic Reconnection

This chapter presents some of the main concepts that are associated to magnetic reconnection and that are extensively used thought the Thesis. Firstly, the definition of the frozen-in concept is given, in order to introduce the concept of magnetic reconnection in the Magneto-Hydro-Dynamics (MHD) theoretical framework (Sect. 2.1). Then, the Ohm's law is derived in Sect. 2.3 and the relative importance of the terms at different scale is discussed. Section 2.4 discusses the origin of the Hall magnetic field. Finite Larmor radius effects focusing on the meandering orbits and the corresponding distribution functions are discussed in Sect. 2.5. These sections are meant to provide all the fundamental notions about the magnetic reconnection process that are necessary to understand the results presented in the following chapters.

2.1 Frozen-In Concept

Even though the MHD approach neglects the kinetic processes at play in the plasma, it still represents an useful theory to understand large scale properties of a plasma, e.g. the frozen-in concept and the magnetic reconnection process at its largest scale.

We consider plasmas that are magnetized, collisionless and fully ionized, composed by protons and electrons. Within the ideal MHD framework, the equation that describes the evolution of the magnetic field reads

$$\frac{\partial \mathbf{B}}{\partial t} = \nabla \times (\mathbf{u} \times \mathbf{B}) \tag{2.1}$$

where $\mathbf{u} = (m_p \mathbf{u}_i + m_e \mathbf{u}_e)/(m_i + m_e)$ is the bulk velocity of the single fluid plasma. Equation (2.1) is obtained combining the Faraday equation with the ideal Ohm's law

G. Cozzani, *Microphysics of Magnetic Reconnection in Near-Earth Space*, Springer Theses, https://doi.org/10.1007/978-3-030-56142-0_2

$$\mathbf{E} + \frac{\mathbf{u} \times \mathbf{B}}{c} = 0 \tag{2.2}$$

In this condition, the plasma is *frozen* to the magnetic field, meaning that if the infinitesimal distance between two fluid element of the plasma $\delta\mathbf{r}$ is tangential to a field line at time t_0 this will still be the case at any further time, i.e.

$$\frac{d}{dt}(\delta\mathbf{r} \times \mathbf{B}) = 0.$$

The frozen-in condition implies that the magnetic field line cannot break and so it introduces crucial constraints regarding the magnetic field topology of the system.

Equation (2.1) implies that the plasma described as a single fluid is frozen-in. However, a similar condition can be obtained for a single species and even for a specific population of a species which is globally frozen-in. This situation takes place, for instance, in the ion diffusion region, where electrons are still coupled to the magnetic field while ions are demagnetized (see Sect. 5.2.1).

Also, when ideal MHD is valid, the magnetic flux Φ_B through a surface $S(t)$ is conserved

$$\frac{d\Phi_B}{dt} = \frac{d}{dt}\left(\int_{S(t)} d\mathbf{S} \cdot \mathbf{B}\right) = 0.$$

The integration is performed over the surface $S(t)$ bounded by the closed curve $\gamma(t)$ which moves with the same velocity field as the plasma. Equation (2.1) can be easily derived considering that the variation $d\mathbf{S}$ can be written as $d\mathbf{S} = \mathbf{u}dt \times d\mathbf{l}$ where $d\mathbf{l}$ is tangential to $\gamma(t)$, and including the ideal Ohm's law and Eq. (2.1). Hence, this property can be seen as equivalent to the frozen-in concept.

When the plasma is frozen to the magnetic field, the motion of the field lines is linked to the motion of the plasma fluid element so that they are both characterized by the same velocity. This characteristic velocity $\mathbf{u}_{E\times B}$, also called the drift velocity, can be obtained by performing the cross product of $\mathbf{B}$ and Eq. (2.2) so that $\mathbf{u}_{E\times B} = c\frac{\mathbf{E}\times\mathbf{B}}{B^2}$.

2.2 Overview of the Magnetic Reconnection Process

This section in meant to provide a short review of some fundamental concepts associated with reconnection that will be useful to understand the next chapters which are focused on the microphysics of the process.

The ideal MHD, with its properties discussed in the previous section, is a theoretical framework that can be used to model the large-scale collisionless plasma dynamics where the fluctuations are characterized by typical scale length L and time scale T such that kinetic effects can be neglected. However, there are some situations in which the characteristic scale is locally decreased down to the scales of the particles scales. This leads to the formations of regions in which ideal MHD is locally violated so that the system can modify its magnetic topology and adjust to a lower energy state, usually by converting energy explosively. The formation of such regions is usually promoted by the presence of current sheets which are regions of strong gradients in the magnetic field. Magnetic reconnection is the process responsible for such energy release and magnetic field topology change.

This prefatory description of the reconnection process already underlines the intrinsic multiscale nature of magnetic reconnection.

The large scale system, which may be modeled by the ideal MHD theory, is affected by the particle-scale processes that take place in regions where the ideal MHD is locally violated. These local and small scale processes eventually lead to the reconfiguration of the global topology and connectivity of field lines, which allows the mixing of regions of plasma that were initially "disconnetted" and which affects the paths of fast particles and heat conduction. For example, the mixing between the solar sourced plasma and the Earth's magnetosphere driven by magnetic reconnection in the near-Earth space has crucial consequences on the terrestrial magnetosphere, as discussed in Sect. 3.1.

In order to introduce the process of magnetic reconnection we consider the simplest 2D configuration and we describe the evolution of the system adopting a MHD (large-scale) point of view (Priest and Forbes 2000, e.g.). Let us consider two adjoining regions characterized by oppositely oriented magnetic field lines (see Fig. 2.1a). Consistently with the changing in the direction of the magnetic field, a current sheet boundary is present. Under enhanced gradients across the boundary driven by the system dynamics, the current sheet becomes thinner and the configuration becomes unstable. Because of the enhanced gradients, a non-ideal region is formed associated to a neutral X-point where the magnetic field vanishes (at least in this simplified picture shown in Fig. 2.1b). Within this non-ideal region Eq. (2.1) is no longer valid. This region has been referred to as the *diffusion region* since the magnetic field can finally slip with respect to the plasma. We also recall that Eq. (2.1) can be seen as the limit for $\eta \rightarrow 0$ of the equation for the evolution of the magnetic field in a *resistive* plasma

$$\frac{\partial \mathbf{B}}{\partial t} = \nabla \times (\mathbf{u} \times \mathbf{B}) + \frac{c^2}{4\pi}\eta \nabla^2 \mathbf{B}$$

where η is the resistive diffusion coefficient. Resistivity provides an example of dissipation and the necessary irreversible dissipation for reconnection to occur.

The processes taking place in the diffusion region lead to the breaking and reconnection of magnetic field lines together with the release of magnetic energy that is converted in kinetic energy of the plasma. In particular, the plasma is accelerated and heated in the outflow region (Fig. 2.1c).

The current MHD description of magnetic reconnection can not provide any detail about the kinetic processes in the diffusion region, but it allows to understand several crucial properties of this process at large scale. In particular,

1. Magnetic reconnection is driven by the presence of strong gradients and thin current sheets;

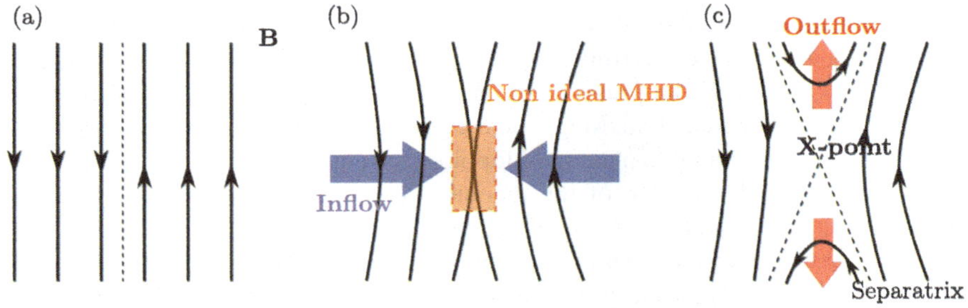

Fig. 2.1 Schematic picture of magnetic field merging. Black lines represent magnetic field lines

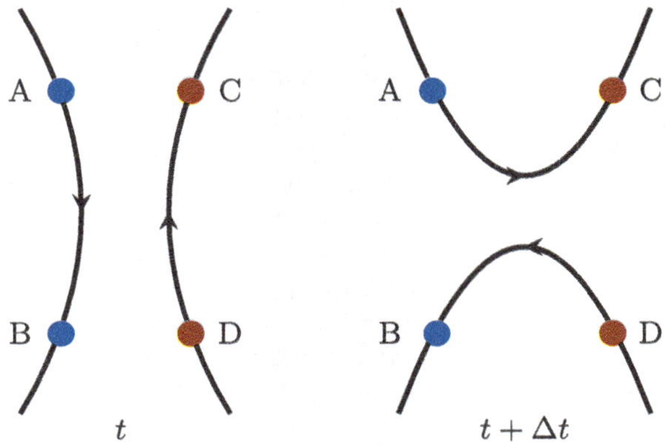

Fig. 2.2 Schematic representation of the change in connectivity operated by the reconnection process. At time t, prior to reconnection, the fluid element A and B (C and D) are frozen to the same field line (left panel). After reconnection (right panel), the connectivity has changed and now A and C (B and D) are connected

2. Magnetic reconnection is a multiscale process; the local non-ideality of the plasma induces global variations in the topology and in the energy partition between fields and the plasma;
3. During reconnection magnetic energy is converted into heat, and kinetic plasma energy;
4. As a consequence of the topology reorganisation, plasma of different regions and source that were initially apart can mix (see Fig. 2.2).

Note that the name *reconnection* is issued by the MHD theory. From the MHD perspective the field lines do break and re-connect, but the MHD model is not describing the processes within the diffusion region where the "breaking" and "reconnecting" take place. In addition, the notion of field lines itself has to be taken with care once we consider non-ideal regions where the frozen-in condition does not hold since it is no longer possible to prove that there is a well defined equation for the evolution of the "field lines" in that context.

2.2.1 Reconnection Rate

The reconnection rate is defined as the amount of the reconnected magnetic flux flowing into or away from the reconnection site (in 2D, this is usually the X-point). It is one of the main parameters that characterise the reconnection process. If the system can be described by ideal MHD the magnetic flux remains constant

$$\frac{\partial \Phi_B}{\partial t} = 0 \tag{2.3}$$

consistently with the frozen-in condition. During reconnection this condition is no longer valid in the diffusion region and the reconnected magnetic flux across the region changes in time. This variation quantifies the speed at which reconnection is proceeding. It can be shown that the reconnection rate $\mathcal{R}$ can be expressed (using Faraday's law and the Stokes theorem) in terms of the electric field in the reconnection site

$$\mathcal{R} \equiv \frac{\partial \Phi_B}{\partial t} = \frac{\partial}{\partial t}\left(\int_S d\mathbf{S}\cdot\mathbf{B}\right) = -c\oint_\gamma d\mathbf{l}\cdot\mathbf{E} \tag{2.4}$$

where S is surface bounded by the closed curve γ.

Now, let us consider a simplified 2D geometry as the one previously adopted (see Fig. 2.1). There is an invariant direction which corresponds to the out of reconnection plane direction (z direction in Fig. 2.3). Also, let us assume that the system is in steady state. Here, the reconnection rate is a measure of the rate at which magnetic flux is transported across the X-point.

We consider the curve γ bounding the surface S defined above has one side passing across the X-point and one side far enough from the reconnection site to be considered as not affected by reconnection (side a and c of the rectangle in Fig. 2.3). We compute the contour integral in Eq. (2.4) by decomposing it along the four segment of γ. The contribution of segment c is zero since the electric field is supposed to be zero there; the joint contribution of segments b and d is zero since the electric field is the same on the two segments (invariance along z). The only contribution is given by segment a so that

$$\mathcal{R}_{2D} = -c\, E_{z,X}\, a \tag{2.5}$$

where $E_{z,X}$ is the out of reconnection plane component of the electric field at the X-point, c is the speed of light and a is the length of the side of the curve γ. Hence, $E_{z,X}$ can be used to estimate the reconnection rate and it is commonly referred to as the reconnection electric field.

Following the approach by Sweet and Parker, we can write the reconnection electric field $E_{z,X}$ in terms of the Alfvén velocity computed in the inflow region. Assuming mass conservation across the X-point we can write

$$\rho_{in} u_{in} \Delta \sim \rho_{out} u_{out} \delta \tag{2.6}$$

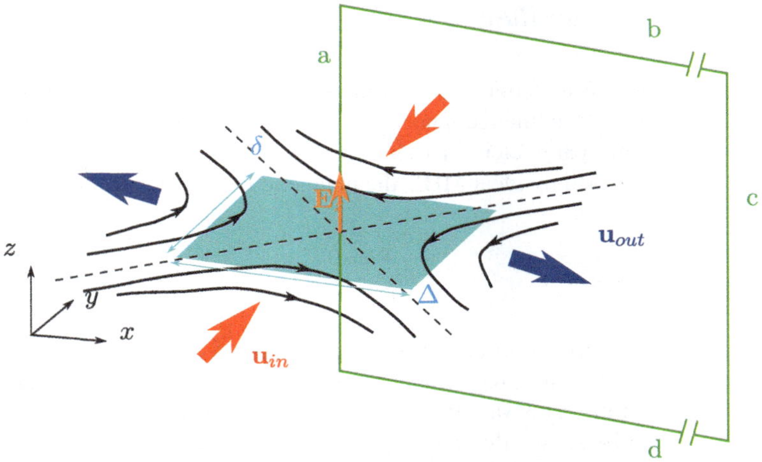

Fig. 2.3 The integration loop is represented by the solid green line. The diffusion region is represented with the light blue rectangle. Reconnection electric field, magnetic field, and in/outflow velocities, are shown in orange, black, and red/blue respectively

where Δ and δ are the characteristic length and width of the non-ideal region (see Fig. 2.3), ρ is the density and u the plasma velocity. The *in* and *out* subscripts correspond to region upstream and downstream of the diffusion region where ideal MHD is valid. For equal density in the inflow and outflow region and considering that the total energy is dominated by magnetic energy in the inflow region and by the plasma kinetic energy in the outflow region, we get

$$\Delta \frac{B_{in}^2}{8\pi} u_{in} \sim \delta \frac{1}{2} \rho_{out} u_{out}^2 u_{out} \tag{2.7}$$

Hence, using Eq. (2.6) we obtain

$$u_{out} \sim \sqrt{\frac{B_{in}^2}{4\pi\rho}} \equiv V_{A,in} \tag{2.8}$$

which corresponds to the Alfvén velocity in the inflow region $V_{A,in}$. In the inflow region, the electric field is still described by the ideal Ohm's law Eq. (2.2) so that

$$E_z = \frac{u_y B_x}{c} = \frac{u_{in} B_{in}}{c} \sim \frac{\delta}{\Delta} \frac{V_{A,in} B_{in}}{c}. \tag{2.9}$$

In conclusion, the reconnection rate depends upon the aspect ratio δ/Δ of the diffusion region which in turn depends on the microphysics at play in the non-ideal region.

The reconnection rate is often presented in an adimensional fashion by dividing it by $B_{in} V_{A,in}/c$. In this way, we obtain that the adimensional reconnection rate $\overline{E}_z$ is

$$\overline{E}_z = \frac{cE_z}{B_{in} V_{A,in}} = \frac{u_{in} B_{in}}{c} \frac{c}{B_{in} V_{A,in}} = \frac{u_{in}}{V_{A,in}}. \tag{2.10}$$

Moreover, since the magnetic flux into and out of the dissipation region is the same, $B_{in} u_{in} = B_{out} u_{out}$ (B_{in} is directed along x and B_{out} along y)

$$\overline{E}_z = \frac{cE_z}{B_{in} V_{A,in}} = \frac{u_{out} B_{out}}{c} \frac{c}{B_{in} V_{A,in}} = \frac{B_{out}}{B_{in}} = \frac{B_y}{B_x}. \tag{2.11}$$

This formulation is important to experimentally compute the reconnection rate. Indeed, spacecraft trajectories do not provide enough information to compute the reconnection rate using Eq. (2.4) and the out-of-reconnection-plane electric field is usually a small quantity affected by large errors. In contrast, the inflow velocity, together with the Alfvén speed and the components of the magnetic field B_y and B_x is usually available, allowing for an estimation of the reconnection rate. However, these measurements need to be handled with care since B_y is usually close to zero in observations (as compared to B_x) especially for asymmetric reconnection. Furthermore, the motion of the X-point has to be determined in order to evaluate u_{in}.

In addition, the ratio B_y/B_x can be written in terms of an *opening angle* θ, which is the angle between the two separatrices

$$\overline{E}_z = \frac{B_y}{B_x} = \tan\theta. \tag{2.12}$$

When talking about the reconnection rate, it is impossible not to mention the so-called "0.1 reconnection rate problem". Indeed, it has been shown that the reconnection rate takes the 0.1 value in myriad of numerical simulations performed in a wide range of conditions, see e.g. Comisso and Bhattacharjee (2016), Cassak et al. (2017b), and references therein.

However, it is still not clear why the reconnection rate should take this specific value. The discussion and the investigation of the "0.1 reconnection rate problem" lie beyond the scope of this Thesis. The reader is referred to publications that approach this long lasting problem (Cassak et al. 2017b; Comisso and Bhattacharjee 2016; Dorelli and Bhattacharjee 2008; Priest and Forbes 2000).

2.3 Generalized Ohm's Law

The electric field in the non-ideal region is referred to as the reconnection electric field. As shown in the previous section, it is proportional to the reconnection rate so that it is also linked to the energy conversion from the magnetic field to the plasma.

The overall MHD viewpoint presented in the previous section is not suitable to provide details about the microphysics sustaining the reconnection electric field. In the diffusion region, where the ideal Ohm's law is no longer valid, the electric field is given by

$$\mathbf{E} + \frac{\mathbf{u} \times \mathbf{B}}{c} = \mathbf{R} \tag{2.13}$$

where $\mathbf{R}$ represents the non-ideal terms.

In a collisional and resistive plasma, the right hand side $\mathbf{R}$ corresponds to a term of the type $\eta\mathbf{j}$, describing Ohmic dissipation. In a collisionless plasma, the resistivity is included only for the sake of mathematical simplicity but it does not correspond to the correct physics at small scales. Indeed, as soon as the system generates strong gradients in the reconnecting layer, other terms come into play in the Ohm's law well before resistive effects would be efficient.

In order to write the right hand side of Eq. (2.13) more explicitly, we consider the Euler equation for both electrons and ions

$$\frac{\partial}{\partial t}(n\mathbf{u}_e) + \nabla \cdot (n\mathbf{u}_e\mathbf{u}_e) = -\frac{1}{m_e}\nabla \cdot \overline{\mathbf{P}_e} - \frac{ne}{m_e}\left(\mathbf{E} + \frac{\mathbf{u}_e \times \mathbf{B}}{c}\right) \tag{2.14}$$

$$\frac{\partial}{\partial t}(n\mathbf{u}_i) + \nabla \cdot (n\mathbf{u}_i\mathbf{u}_i) = -\frac{1}{m_i}\nabla \cdot \overline{\mathbf{P}_i} + \frac{ne}{m_i}\left(\mathbf{E} + \frac{\mathbf{u}_i \times \mathbf{B}}{c}\right) \tag{2.15}$$

where we have assumed quasi-neutrality $n_e \simeq n_i \simeq n$. Since $\mathbf{j} = ne(\mathbf{u}_i - \mathbf{u}_e)$, the left hand side of Eq. (2.14) can be written as

$$\frac{\partial}{\partial t}(n\mathbf{u}_e) + \nabla \cdot (n\mathbf{u}_e\mathbf{u}_e) = \frac{\partial}{\partial t}\left(n\mathbf{u}_i - \frac{\mathbf{j}}{e}\right) + \nabla \cdot (n\mathbf{u}_i\mathbf{u}_i) - \frac{1}{e}\nabla \cdot (\mathbf{u}_i\mathbf{j}) - \frac{1}{e}\nabla \cdot (\mathbf{j}\mathbf{u}_i) + \frac{1}{ne^2}\nabla \cdot (\mathbf{j}\mathbf{j}). \tag{2.16}$$

Then, we subtract Eq. (2.14) from (2.15) and we finally obtain

$$\mathbf{E}\left(1 + \frac{m_e}{m_i}\right) + \frac{\mathbf{u}_e \times \mathbf{B}}{c} + \frac{m_e}{m_i}\frac{\mathbf{u}_i \times \mathbf{B}}{c} = \frac{1}{ne}\nabla \cdot \left(\frac{m_e}{m_i}\overline{\mathbf{P}_i} - \overline{\mathbf{P}_e}\right) + \\ + \frac{m_e}{ne^2}\nabla \cdot \left(\mathbf{j}\mathbf{u}_i + \mathbf{u}_i\mathbf{j} - \frac{\mathbf{j}\mathbf{j}}{ne}\right) + \frac{m_e}{ne^2}\frac{\partial \mathbf{j}}{\partial t}. \tag{2.17}$$

No approximations have been used yet, but we can simplify the above equation by considering that $1 + m_e/m_i \approx 1$ and that usually $(m_e/m_i)|\mathbf{u}_i| \ll |\mathbf{u}_e|$. Hence, we obtain the generalised Ohm's law

$$\mathbf{E} + \frac{\mathbf{u}_e \times \mathbf{B}}{c} = \frac{1}{ne}\nabla \cdot \left(\frac{m_e}{m_i}\overline{\mathbf{P}_i} - \overline{\mathbf{P}_e}\right) + \frac{m_e}{ne^2}\nabla \cdot \left(\mathbf{j}\mathbf{u}_i + \mathbf{u}_i\mathbf{j} - \frac{\mathbf{j}\mathbf{j}}{ne}\right) + \frac{m_e}{ne^2}\frac{\partial \mathbf{j}}{\partial t}. \tag{2.18}$$

The left hand side can be also written in terms of the ion velocity and we obtain

$$\mathbf{E}+\frac{\mathbf{u}_i\times\mathbf{B}}{c}=\frac{\mathbf{j}\times\mathbf{B}}{nec}+\frac{1}{ne}\nabla\cdot\left(\frac{m_e}{m_i}\overline{\mathbf{P}_i}-\overline{\mathbf{P}_e}\right)+\frac{m_e}{ne^2}\nabla\cdot\left(\mathbf{j}\mathbf{u}_i+\mathbf{u}_i\mathbf{j}-\frac{\mathbf{j}\mathbf{j}}{ne}\right)+\frac{m_e}{ne^2}\frac{\partial\mathbf{j}}{\partial t}. \tag{2.19}$$

In this formulation, the so-called Hall term $\propto \mathbf{j}\times\mathbf{B}$ is explicitly present. In addition, the term depending on the species pressure is called the pressure term while the term depending upon $\mathbf{u}_i$ and $\mathbf{j}$, as well as on the temporal derivative of $\mathbf{j}$, and proportional to m_e is called the electron inertia term. In general, $\frac{m_e}{m_i}|\nabla\cdot\overline{\mathbf{P}_i}|\ll|\nabla\cdot\overline{\mathbf{P}_e}|$ so that the ion pressure tensor contribution may be neglected.

We compare the different terms in order to understand at which scale each of them is actually playing a role. This comparison underlines the multiscale nature of the reconnection process. It is straightforward to see that the two term on the left hand side of Eq. (2.19) are of the same order while the following ordering is valid for the other terms

$$\left|\frac{\mathbf{j}\times\mathbf{B}/nec}{\mathbf{u}_i\times\mathbf{B}/c}\right|\sim\frac{d_i^2}{L^2} \tag{2.20}$$

$$\left|\frac{\nabla\cdot\overline{\mathbf{P}_e}/ne}{\mathbf{u}_i\times\mathbf{B}/c}\right|\sim\beta\frac{d_i^2}{L^2} \tag{2.21}$$

$$\left|\frac{\frac{m_e}{ne^2}\frac{\partial\mathbf{j}}{\partial t}}{\mathbf{u}_i\times\mathbf{B}/c}\right|\sim\left|\frac{\frac{m_e}{ne^2}\nabla\cdot\left(\mathbf{j}\mathbf{u}_i+\mathbf{u}_i\mathbf{j}-\frac{\mathbf{j}\mathbf{j}}{ne}\right)}{\mathbf{u}_i\times\mathbf{B}/c}\right|\sim\frac{d_e^2}{L^2} \tag{2.22}$$

where L corresponds to the characteristic scale of interest. The Hall term plays a role at the ion scales d_i while the inertia term becomes crucial at the electron scale d_e. The relative importance of the pressure term depends also on the β parameter, the ratio of kinetic to magnetic pressure.

All terms on the right hand side of Eq. (2.19) (and Eq. (2.18)) contribute to the non-ideal $\mathbf{R}$ term on the right hand side of Eq. (2.13). However, not all of them guarantee the demagnetization of both species as required by the reconnection process to occur. In particular, the Hall term is sufficient to demagnetize the ions but not the electrons. Indeed, the Ohm's law with the Hall term as the only non-ideal contribution reads

$$\mathbf{E}=-\frac{\mathbf{u}_i\times\mathbf{B}}{c}+\frac{\mathbf{j}\times\mathbf{B}}{nec}=-\frac{\mathbf{u}_e\times\mathbf{B}}{c} \tag{2.23}$$

which corresponds to an ideal Ohm's law for electrons.

If we take into account the pressure term we obtain the following Ohm's law

$$\mathbf{E}=-\frac{\mathbf{u}_e\times\mathbf{B}}{c}-\frac{1}{ne}\nabla\cdot\overline{\mathbf{P}_e} \tag{2.24}$$

which in principle allows for the electrons to be demagnetized. However, by taking the curl of Eq. (2.24) and using Faraday's law we get

$$\frac{\partial \mathbf{B}}{\partial t} = \nabla \times (\mathbf{u}_e \times \mathbf{B}) + \frac{c}{e}\nabla \times \left(\frac{\nabla \cdot \overline{\mathbf{P}_e}}{n}\right). \tag{2.25}$$

If the pressure is a scalar or a diagonal tensor and the plasma is incompressible, the "diffusive" term is zero the electrons cannot be demagnetized.

Finally, the electron inertia term is sufficient to break the frozen-in condition and it is in general considered to play a major role at electron scales.

2.4 Hall's Quadrupole

The previous section shows that at the ion scales the Hall term becomes non-negligible and it allows for the decoupling of electron and ion flows. The fact that charge carriers of different sign now move differently allows for an out-of-plane component of the magnetic field to develop. This out-of-reconnection-plane magnetic field has a quadrupolar structure and it represents one of the most simple to observe and unambiguous signatures of ongoing reconnection (Sonnerup 1979). The quadrupolar magnetic field has been routinely observed in numerical simulations (Biskamp et al. 1997; Pritchett 2001; Shay and Drake 1998; Shay et al. 2006, 2001; Wang et al. 2000), near-Earth space (Eastwood et al. 2010; Mozer et al. 2002) and laboratory experiments (Ren et al. 2005) (Figs. 2.4 and 2.5).

In order to explain the origin of the quadrupolar pattern, we report the explanation given by Uzdensky and Kulsrud (2006). We consider a 2D reconnecting current layer in the Hall-MHD regime with slab geometry. The xy plane is the so-called reconnection plane and the system is invariant along the direction z. In particular, we consider the ion diffusion region where the ions are demagnetized from the magnetic field and they can be consider as a motionless neutralizing background. We assume that the ion density is uniform across the current layer. Since quasi-neutrality is assumed, the electron density is also uniform. We also assume that the reconnection layer is in a quasi-steady state and that there is no guide magnetic field (i.e. no initial magnetic field along the z direction).

We consider a flux tube that is moving towards the X-point, as shown in Fig. 2.6a. The magnetic field strength decreases going towards the center of the reconnecting layer and the volume of the tube needs to increase in order to guarantee flux conservation. Since the electron are still frozen-in, this leads to a drop in the electron density. However, the ions are not influenced by this dynamics and their density stays uniform. This induces an electric field which accelerates electron from the sides of the flux tube towards the center in order to re-establish quasi neutrality (see Fig. 2.6b).

In the outflow region, the direction of the electron flow reverses. The magnetic flux tube is ejected from the X-point along the x direction and the magnetic strength increases in the center of the flux tube so that the electrons move outward in order to maintain quasi-neutrality. The final pattern of the electron flow is shown in Fig. 2.6c. Therefore, the electron-carried current density determines the presence

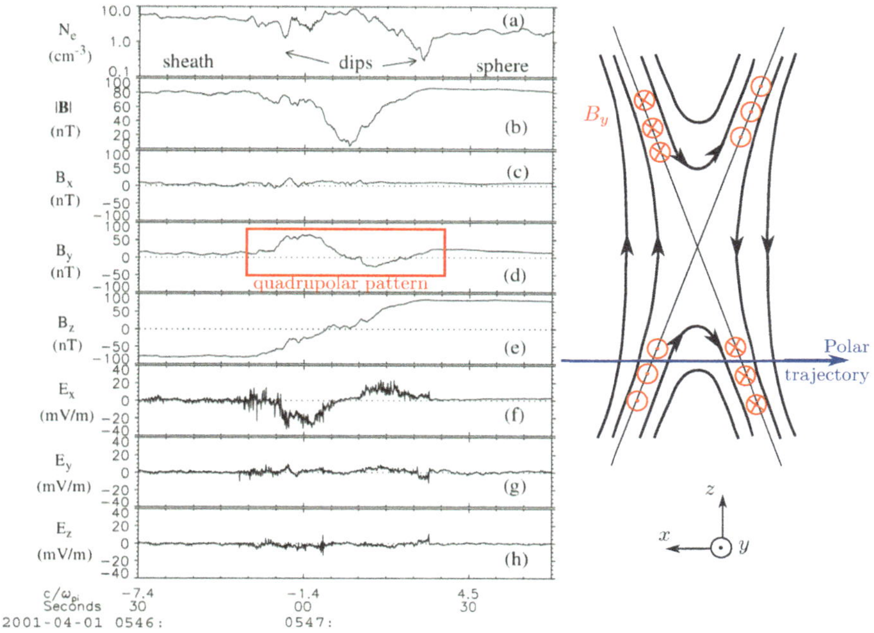

Fig. 2.4 The quadrupolar magnetic field (panel (d)) observed at the magnetopause as reported by Mozer et al. (2002). The data are from the Polar satellite and the spacecraft trajectory across the Hall region is sketched on the left. Reprinted figure with permission from F. S. Mozer, S. D. Bale, and T. D. Phan, Phys. Rev. Lett. **89**, 015002 (2002). © 2002 by the American Physical Society

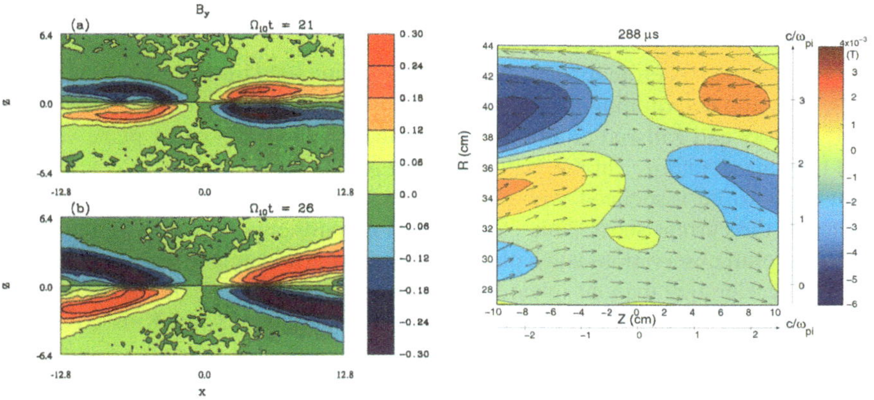

Fig. 2.5 (left) The quadrupolar magnetic field during 2D PIC simulations of magnetic reconnection. Adapted from (Pritchett 2001). © 2001 by the American Geophysical Union. (right) The quadrupolar magnetic field detected in the MRX (Magnetic Reconnection Experiment). Reprinted figure with permission from Y. Ren, M. Yamada, S. Gerhardt, H. Ji, R. Kulsrud, and A. Kuritsyn, Phys. Rev. Lett. **95**, 055003 (2005) (Ren et al. 2005). © 2005 by the American Physical Society

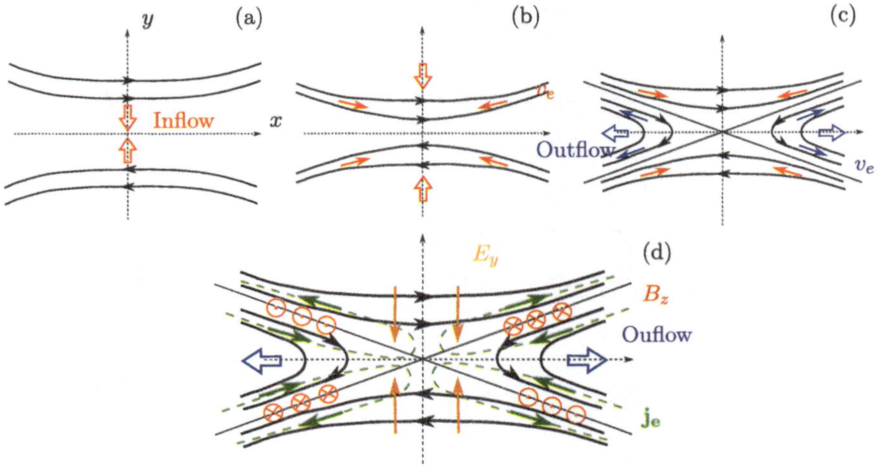

Fig. 2.6 Schematic representation of the explanation for the out-of-plane quadrupolar magnetic field

of a quadrupolar magnetic field concentrated along the separatrices as shown in Fig. 2.6d. Note that the sign pattern is independent of the direction of the magnetic field in the xy plane since the quadrupolar structure is determined by the electron flow only. Also, in this scenario, the electrons are always magnetized and the electron diffusion region is not taken into account.

Another explanation for the quadrupolar magnetic field has been given by Mandt et al. (1994). This explanation invokes the differential drag of the magnetic field lines by a non uniform out-of-plane electron velocity $v_{e,z}$. Being $v_{e,z}$ larger at the neutral line (where $B = 0$ before reconnection), the magnetic field lines are stretched out in the z direction determining the quadrupolar pattern (Fig. 2.7).

Uzdensky and Kulsrud (2006) investigate the origin of the out-of-plane velocity and they argue that $v_{e,z}$ in the Hall region corresponds to the drift velocity $\mathbf{E} \times \mathbf{B}$ which has to be associated with a bipolar electric field E_y directed towards the center of the current sheet. This electric field is routinely observed in simulations and spacecraft observations and it plays a role in the ion dynamics in the current sheet, namely accelerating the ions towards the neutral line. In addition, it has been shown that, at least in the analytic X-line model adopted by Uzdensky and Kulsrud (2006), the bipolar E_y is sensibly larger that the out-of-plane electric field (or reconnection electric field) and this behaviour is recovered in spacecraft observations (Burch et al. 2016a; Mozer et al. 2002).

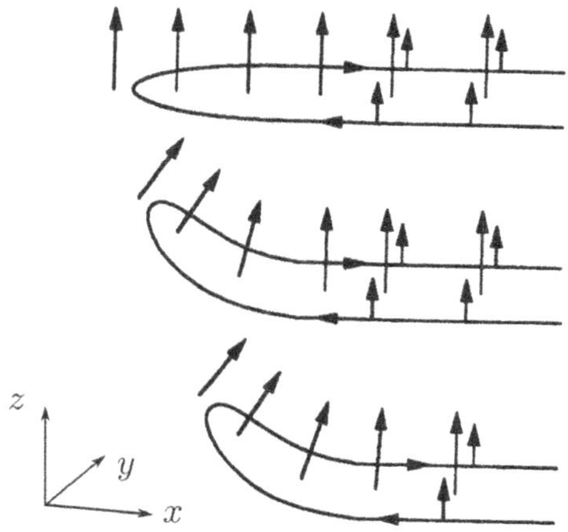

Fig. 2.7 A schematic representation of the stretching of the reconnected magnetic field lines in the z direction caused by the non uniform $v_{e,z}$ across the current sheet. The loop represents the magnetic field line initially lying the xy plane. The arrows directed along z represent the electron velocity which is larger in the center of the current sheet. Adapted from Mandt et al. (1994). © 1994 by the American Geophysical Union

2.5 Finite Larmor Radius Effects: Meandering Orbits

One of the first model of particle dynamics within a current sheet where the sheared magnetic field changes direction is the Speiser analytic model (Speiser 1965). Speiser (1965) developed a single particle model that is valid in a current sheet where the magnetic field is approximated by the linear function $-(x/\delta)\hat{y}$, where δ is the half thickness of the current sheet, while the electric field is constant and directed along x (Fig. 2.8). In these conditions the Larmor radii of the particles have to be considered as finite.

Once the charged particles enter the current sheet, they start to execute meandering orbits performing oscillations within the sheet and they are accelerated along the sheet before being finally ejected (Fig. 2.8). It is seen in (Speiser 1965) that the number of oscillations in the current sheet is proportional to the square root of the mass and so electrons will execute about 1/40 of the number of proton oscillations before being ejected. Moreover, as indicated in Fig. 2.8, the magnetic field has a small component perpendicular to the sheet which is the main driver for the ejection process of the particles.

The Speiser meandering orbits have recently gained popularity since they provide the explanation for the generation of the so-called *crescent*-shaped distribution functions observed by the MMS spacecraft in proximity of the electron diffusion region at both magnetopause and magnetotail (Burch et al. 2016a; Torbert et al. 2018). This type of distribution has been observed also for ions (Dargent et al. 2019), one of the first example has been reported by Frank et al. (1994) in the magnetotail.

Electron crescent-shaped distributions close to the reconnection site were predicted by Hesse et al. (2014) before the MMS mission was launched. Bessho et al. (2016) developed a one-dimensional models for crescent distributions at the magnetopause in which the magnetic field is varying linearly across the boundary and the electric field is zero on one side and varies linearly on the other. This asymmetry

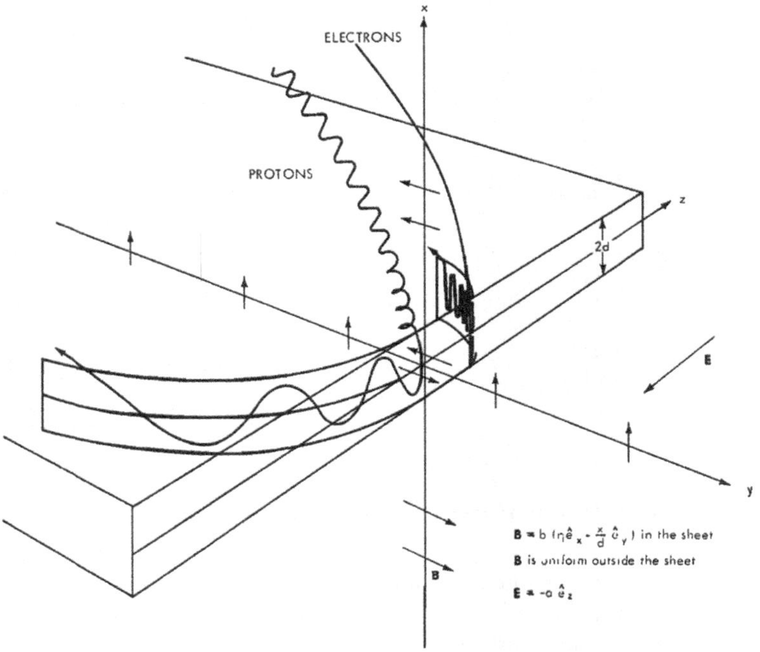

Fig. 2.8 Ion and electron orbit within the current sheet. The magnetic field is $\mathbf{B} = b\hat{x} - x/\delta\hat{y}$ and the electric field is $\mathbf{E} = -a\hat{z}$. Adapted from Speiser (1965). © 1965 by the American Geophysical Union

reflects the asymmetric reconnection at the magnetopause which will be discussed in Sect. 3.2. Another one-dimensional model was independently developed by Shay et al. (2016). In this case both electric and magnetic field are asymmetric and they are zero on one side of the boundary and linearly increasing on the other. Both models recover the crescent-shaped distributions. Note that in the model proposed by Shay et al. (2016) the position of the crescent depends on the free parameter $V_{EB} = cE/B$ which is proportional to the absolute value of the drift velocity. The same free parameter is found by Bessho et al. (2016) (see Fig. 2.9).

Despite these models, the origin of the crescent-shaped distribution functions in still debated. In particular, Lapenta et al. (2017) showed that the electric field is not necessary to obtain the meandering orbits. In contrast, the only necessary condition is a steep reversal of the magnetic field. Since the electric field is not taken into account, the parameter V_{EB} does not appear.

The model developed by Lapenta et al. (2017) is based on the adiabatic Hamiltonian method. It allows both magnetic and electric fields to vary linearly and asymmetry is not necessary. Since both asymmetric and symmetric configurations provide crescent-shaped distribution functions, the model predicts the presence of the meandering orbits also in a magnetotail-like (symmetric) configuration. Indeed, crescent-

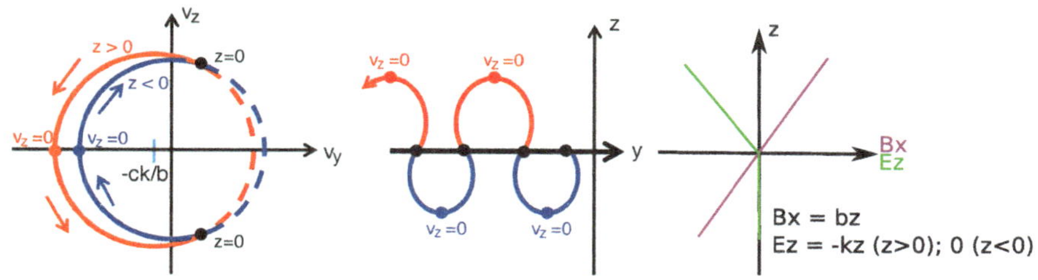

Fig. 2.9 Schematic representation of the crescent-shaped distribution function in the velocity space and the corresponding orbit in the xz plane. The reconnection plane is the xz plane and the z direction is normal to the current sheet. The parameter $-ck/b$ corresponds to V_{EB}. Adapted from Bessho et al. (2016). © 2016 by the American Geophysical Union

shaped distribution functions are observed by MMS in the magnetotail (Torbert et al. 2018).

In addition, this model underlines the fact that electron crescent-shaped distribution functions easily develop near the Electron Diffusion Region because of the extreme thinning of the current sheet near the X-point but they can be observed in any other region in which the magnetic field gradient is strong enough to induce meandering orbits. This point has been confirmed by observations. For instance, crescent-shaped distribution functions are observed at flank magnetopause where they are generated by electron gyromotion at thin electron-scale magnetic boundaries (Tang et al. 2019).

2.6 Standard Picture of 2D Magnetic Reconnection

This section is meant to summarize the previous sections providing a complete picture of the region which is affected by the reconnection process, i.e. the region including the diffusion region, outflows and inflows. The approach of this section is relatively qualitative since the main purpose is to provide elements to better understand the spacecraft data analysis presented in Chap. 7 and the simulations presented in Chap. 8. The properties that are discussed in this section correspond to specific signatures in the observations which allow to identify ongoing reconnection and the ion diffusion region.

This description is limited by the fact that we consider a 2D system which is invariant over translation along the out-of-reconnection-plane direction. This approximation is adopted throughout the Thesis since the spacecraft observations suggest that the analyzed system is rather 2D (as discussed in Chap. 7) and the simulations presented in Chap. 8 are 2D simulations.

As discussed in the previous sections, magnetic reconnection is a multiscale process. In general, we can distinguish three regions: an external region in which the plasma is frozen to the magnetic field, an Ion Diffusion Region (IDR) where elec-

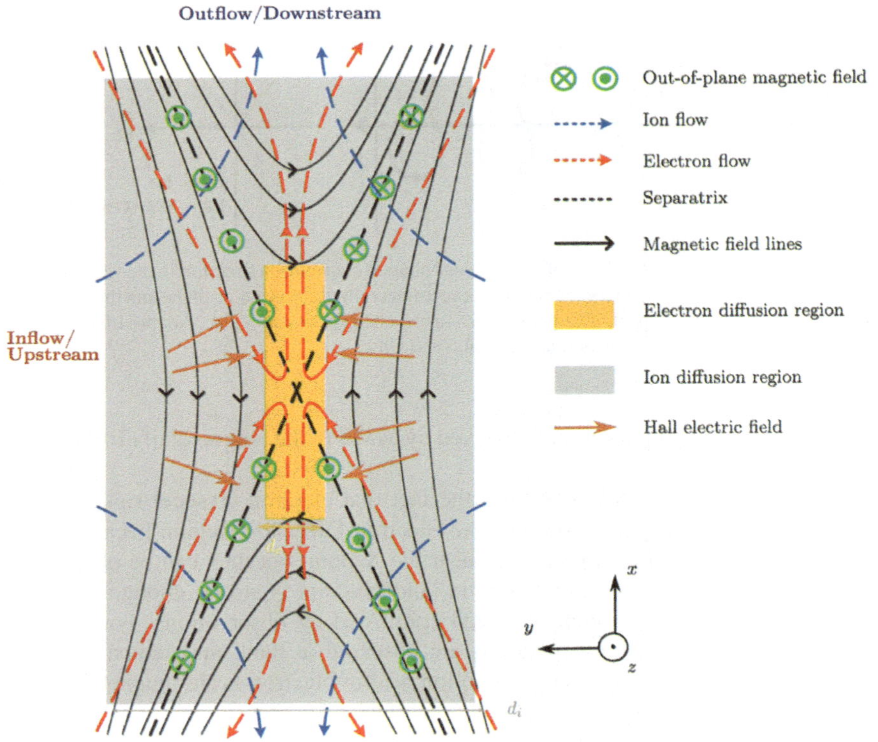

Fig. 2.10 Schematic diagram of the diffusion region. Ions are decoupled from the electrons and from the magnetic field in the ion diffusion region. The Hall magnetic and electric field patterns are present in the IDR. Electrons are demagnetized in the electron diffusion region. Reprinted from M. Yamada, J. Yoo, J. Jara-Almonte, W. Daughton, H. Ji, R. M. Kulsrud, and C. E. Myers, Phys. Plasmas **22**, 056501 (2015) (Yamada et al. 2015), with the permission of AIP Publishing

trons are still frozen-in and an Electron Diffusion Region (EDR), embedded into the IDR, where both electron and ions are demagnetized and the breaking of the field lines takes place (see Fig. 2.10). The typical dimension of each region depends on the terms of the Ohm's law which determines the demagnetization of the species. More precisely, the characteristic scale of the IDR is the ion inertial length d_i (the characteristic scale length of the Hall term) while the characteristic scale of the EDR is the electron inertial length d_e.

In the inflow (upstream) region, the ideal MHD is approximately valid. During reconnection, a flow is directed towards the current sheet, convecting the magnetic field inward. Since Eq. (2.2) holds, an out-of-plane electric field $E_z = u_y B_x$ is present.

Once the inflowing field lines enter into the IDR, they decouple from the ions while the electrons stay frozen-in. This behaviour is marked by the presence of the so-called Hall currents (see red paths in Fig. 2.10) which sustain the out-of-plane quadrupolar Hall magnetic field. As discussed in Sect. 2.4, a Hall electric field is also present. It is directed along the y direction pointing towards the center of the current sheet. In the ion diffusion region, the ions have very little bulk motion and they perform Speiser orbits. However, the magnetized electrons continue to inflow, so that there is still an out-of-plane electric field, mainly sustained by the Hall term $\propto \mathbf{j} \times \mathbf{B}$ (see Eq. (2.19)).

At wavelengths comparable to d_e the electrons also become demagnetized. It has been proposed that the out-of-plane electric field is associated with the divergence of the off-diagonal elements of the electron pressure tensor (Hesse et al. 1999, 2018). Nevertheless, the role of the electron inertia term is still under debate. The EDR will be discussed in Chap. 6, this section provides only a qualitative picture of the diffusion region. In particular, the only characteristic of the EDR that is taken into consideration at this stage is that the X-point, where the magnetic field lines change topology, is supposedly located at the middle of the EDR. In symmetric reconnection, the stagnation point, the location where the inflow velocity goes to zero, is also at the center of the dissipation region. As discussed in Sect. 3.2, this picture is modified for asymmetric reconnection and one of the consequences is that X-point and stagnation point are not co-located.

In the diffusion region the electrons are accelerated out of the reconnection plane by the reconnection electric field E_z and they gyrate around the reconnected magnetic field B_y. This leads to the formation of the outflow directed along the x direction. In particular, the electrons are ejected from the EDR at the electron Alfvén speed forming the electron reconnection jet. The ions undergo a similar dynamics and they exit the ion diffusion region at the ion Alfvén speed. The electron outflow is mainly concentrated along the separatrices, i.e. the lines forming the boundary between the upstream plasma and the downstream plasma which intersects at the X-line (see Fig. 2.10). Outside of the EDR (but still in the IDR), the electrons are decelerated so that the two species exit the IDR at the same speed, entering the downstream region where they are frozen to the reconnected field lines.

Where and how this "re-magnetization" takes place is still under debate and it is strongly connected to the determination of the extent of the dissipation regions in the outflow direction. The reconnection rate is found to be 0.1 in a myriad of different conditions (see Sect. 2.2.1) and since the reconnection rate is proportional to the aspect ratio of the diffusion region, the extent of EDR in the outflow region is expected to be $\sim$ 10 d_e (being the width of the order of $\sim$1 d_e).

However, simulations (Karimabadi et al. 2007; Nakamura et al. 2016; Shay et al. 2007) revealed the existence of an extended electron diffusion region capable to maintain fast reconnection. In particular, these simulations show that the inner EDR is connected to a super-Alfvénic electron jet where electrons are demagnetized that extends at least for 15 d_i from the X-line in the outflow direction. These results were confirmed and even exceeded by the observation of a super-Alfvénic electron outflow jet at a distance of at least 60 d_i downstream of an active X-line (Phan et al. 2007).

References

N. Bessho, L.-J. Chen, M. Hesse, Geophys. Res. Lett. **43**, 1828 (2016)
D. Biskamp, E. Schwarz, J.F. Drake, Phys. Plasmas **4**, 1002 (1997)
J.L. Burch et al. Science **352**, aaf2939 (2016)
P.A. Cassak, Y.-H. Liu, M.A. Shay, J. Plasma Phys. **83**, 715830501 (2017)
L. Comisso, A. Bhattacharjee, J. Plasma Phys. **82**, 595820601 (2016)
J. Dargent, N. Aunai, B. Lavraud, S. Toledo-Redondo, F. Califano, J. Geophys. Res. **124**, 2497 (2019)
J.C. Dorelli, A. Bhattacharjee, Phys. Plasmas **15**, 056504 (2008)
J.P. Eastwood, T.D. Phan, M. Øieroset, M.A. Shay, J. Geophys. Res. **115**, A08215 (2010)
L.A. Frank, W.R. Paterson, M.G. Kivelson, J. Geophys. Res. **99**, 14877 (1994)
M. Hesse, K. Schindler, J. Birn, M. Kuznetsova, Phys. Plasmas **6**, 1781 (1999)
M. Hesse, N. Aunai, D. Sibeck, J. Birn, Geophys. Res. Lett. **41**, 8673 (2014)
M. Hesse et al., Phys. Plasmas **25**, 032901 (2018)
H. Karimabadi, W. Daughton, J. Scudder, Geophys. Res. Lett. **34**, L13104 (2007)
G. Lapenta et al., J. Geophys. Res. **122**, 2024 (2017)
M.E. Mandt, R.E. Denton, J.F. Drake, Geophys. Res. Lett. **21**, 73 (1994)
F.S. Mozer, S.D. Bale, T.D. Phan, Phys. Rev. Lett. **89**, 015002 (2002)
T. Nakamura, R. Nakamura, H. Haseagawa, Ann. Geophys. **34**, 357 (2016)
B.U. Ö. Sonnerup, in ed. by L.J. Lanzerotti, C.F. Kennel, E.N.Parker. *Solar System Plasma Physics* (North-Holland Publishing Co., Amsterdam, 1979), p. 45
T.D. Phan, J.F. Drake, M.A. Shay, F.S. Mozer, J.P. Eastwood, Phys. Rev. Lett. **99**, 255002 (2007)
E.R. Priest, T. Forbes, *Magnetic Reconnection. MHD Theory and Applications* (Cambridge University Press, Cambridge, 2000)
P.L. Pritchett, J. Geophys. Res. **106**, 25961 (2001)
Y. Ren, M. Yamada, S. Gerhardt, H. Ji, R. Kulsrud, A. Kuritsyn, Phys. Rev. Lett. **95**, 055003 (2005)
M.A. Shay, J.F. Drake, Geophys. Res. Lett. **25**, 3759 (1998)
M.A. Shay, J.F. Drake, R.E. Denton, D. Biskamp, J. Geophys. Res. **103**, 9165 (1998)
M.A. Shay, J.F. Drake, B.N. Rogers, R.E. Denton, J. Geophys. Res. **106**, 3759 (2001)
M.A. Shay, J.F. Drake, M. Swisdak, Phys. Rev. Lett. **99**, 155002 (2007)
M.A. Shay et al., Geophys. Res. Lett. **43**, 4145 (2016)
T.W. Speiser, J. Geophys. Res. **70**, 4219 (1965)
B.-B. Tang et al., Geophys. Res. Lett. **46**, 3024 (2019)
R.B. Torbert et al., Science **362**, 1391 (2018)
D.A. Uzdensky, R.M. Kulsrud, Phys. Plasmas **13**, 062305 (2006)
X. Wang, A. Bhattacharjee, Z.W. Ma, J. Geophys. Res. **105**, 27633 (2000)
M. Yamada, J. Yoo, J. Jara-Almonte, W. Daughton, H. Ji, R.M. Kulsrud, C.E. Myers, Phys. Plasmas **22**, 056501 (2015)

Chapter 3
Magnetic Reconnection at the Earth's Magnetosphere

This chapter provides information about the physical context in which the magnetic reconnection process is studied in this Thesis, namely the near-Earth space, and in particular the magnetopause current sheet. The satellite observations that are discussed in Chap. 7 are focused on an Electron Diffusion Region encounter during magnetic reconnection at the subsolar magnetopause while the simulations analyzed in Chap. 9 presents a case of 2D symmetric magnetic reconnection that corresponds to the magnetotail configuration. The magnetopause and the magnetotail reconnection processes are different in many ways. For instance, magnetopause reconnection directly mediates the interaction between the terrestrial and the interplanetary environment while magnetotail reconnection takes place between field lines of the same source (the Earth's magnetic field). This difference implies that magnetopause reconnection is asymmetric (in terms of magnetic field strength, number density and temperature) while magnetotail reconnection is largely symmetric. The asymmetry strongly affects the geometry and the dynamics of the reconnection site. In addition, the continuous interaction with the interplanetary magnetic field (IMF) and the solar wind implies that the global reconnection rate and the total reconnected flux are influenced by external conditions.

For all these reasons, this chapter provides an overview of the Earth's magnetosphere before focusing on the properties of magnetopause reconnection.

3.1 Overview of the Earth's Magnetosphere

The Earth's magnetosphere is the terrestrial magnetic environment. The unperturbed terrestrial magnetic field would be approximately a magnetic dipole. In reality, the magnetosphere is constantly interacting with the interplanetary magnetic field and,

G. Cozzani, *Microphysics of Magnetic Reconnection in Near-Earth Space*,
Springer Theses, https://doi.org/10.1007/978-3-030-56142-0_3

in particular, with the solar wind plasma which modify the standard dipole. Indeed, particles escaping from the surface of the Sun give rise to the solar wind, a highly ionized and magnetized plasma mainly formed by ions and electrons which convects the interplanetary magnetic field into the interplanetary space.

As the solar wind expands and arrives in proximity of the terrestrial magnetic environment, it is deflected by the magnetospheric magnetic field. Since the solar wind propagates faster than both the magnetosonic and the Alfvén speeds, a collisionless shock – the so-called bow shock – self-consistently forms upstream of the magnetosphere. At the shock, the plasma flow strongly decreases while the plasma density and temperature are increased, together with the magnetic field strength (in general, since the solar wind is an ultrasonic and ultralfvénic plasma, the physical quantities are increased or decreased by a factor of 4). The characteristic width of the bow shock is usually comparable to the local ion gyroradius.

The region between the bow shock and the magnetosphere is composed by the shocked solar wind downstream of the shock and it is called magnetosheath.

The magnetopause is the boundary between the magnetosheath and the Earth's magnetosphere and it is the site where reconnection between field lines advected by the solar wind and those of the magnetosphere can occur. The magnetopause layer is characterized by strong currents due to the spatial variation of the magnetic field from the magnetosheath to the magnetosphere. The magnetic field of the magnetosphere is directed northward while the direction of the IMF may change depending on the global dynamics of the solar wind. The direction of the IMF strongly affects the shape of the whole magnetosphere, as it can be seen in Fig. 3.1.

When the IMF is directed southward, the conditions enabling magnetic reconnection (namely the presence of a thin current sheet) are fulfilled at the subsolar point and magnetic reconnection takes place at low latitude. The reconnection site is marked by the blue oval on the right side of Fig. 3.1a. The reconnected field lines are then advected tailward by the solar wind. The solar wind stretches the magnetosphere on the night side so that a current sheet is present. At this site, the conditions for magnetic reconnection are fulfilled again leading to magnetotail reconnection (occurring in the blue oval on the left side of Fig. 3.1a). The open field lines are reconnected so that there are newly closed terrestrial fields lines moving towards the Earth and new solar wind lines moving away from Earth. This reconnection mediated mechanism that couples the solar wind and the magnetosphere sets up a global plasma dynamics which is usually called the Dungey cycle (Dungey 1961).

The situation is different when the IMF is directed northward. At the subsolar magnetopause the magnetic field maintains the same direction across the boundary while a magnetic field inversion is usually observed at the cusps (the regions of the magnetosphere where the Earth's magnetic field fans out from the magnetic poles, see Fig. 3.1b). In this case, magnetotail reconnection is strongly inhibited because the dynamics on the dayside does not allow for the formation of a thin current sheet on the nightside.

This overview of the Dungey cycle gives an idea of the fact that the solar wind-Earth's magnetosphere is a highly complex coupled system. The position of the magnetopause changes depending on the solar wind pressure that compresses it and

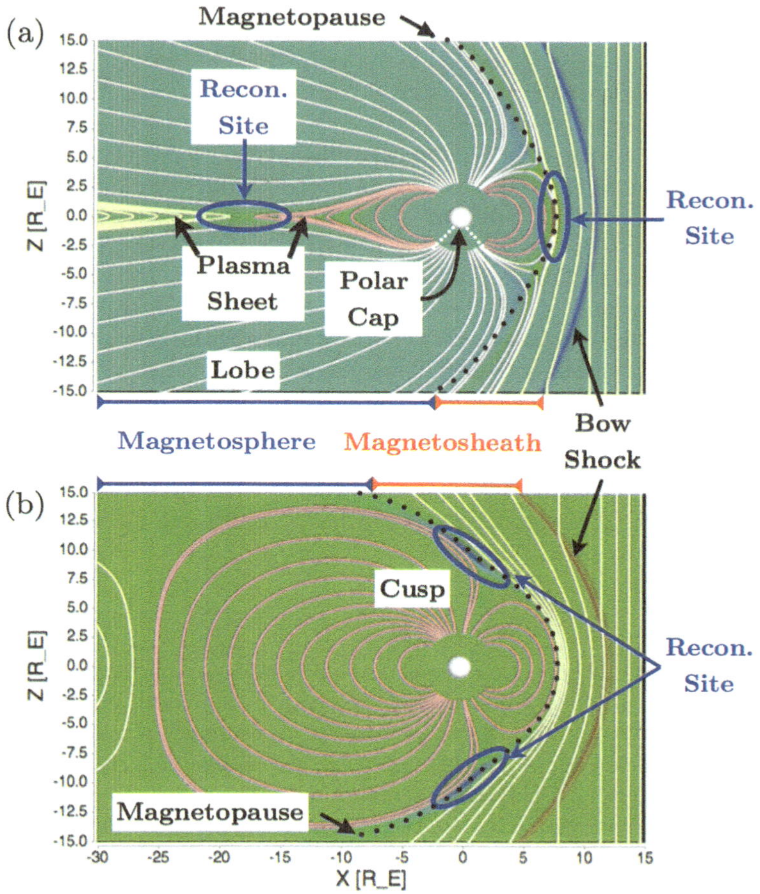

Fig. 3.1 The Earth's magnetosphere with **a** southward IMF and **b** northward IMF obtained with simulations with no dipole tilt. The white lines are the magnetic field lines. The Earth is the white circle and the Sun is to the right (out of picture). The background color represents the out-of-plane current density (blue out of the page and red into the page). Reprinted from Komar (2015) with permission from C. M. Komar

the physical parameters at the magnetopause depend on the solar wind conditions. In addition, the solar wind pressure variations determine a inward and outward motion of the magnetopause. For these reasons, the reconnection site at the magnetopause is not stable and it is not trivial to predict its location. The next section is focused on the properties of magnetopause reconnection at the dayside magnetosphere.

3.2 Properties of Dayside Magnetopause Reconnection

Magnetopause reconnection can be seen as a driven process since it is induced and maintained because the solar wind is compressing the subsolar magnetosphere. In addition, since the solar wind is constantly flowing and its parameters are rapidly changing, one can say that reconnection is always taking place *somewhere* at the magnetopause. However, it is difficult to predict the exact location of the reconnection site. As discussed in the previous section, the location at which magnetopause reconnection is taking place depends on the orientation of the IMF. In particular, it may occurs in proximity of the subsolar point or at the cusps. The real picture is more complicated that the one presented in Fig. 3.1 notably because the dipole tilt has to be taken into account, as well as the possible presence of a $B_{IMF,y} \neq 0$, namely a non-zero component of the IMF magnetic field along the dawn-dusk direction. If reconnection takes place with $B_{IMF,y} \neq 0$ it is called component reconnection since only one component of the magnetic field is actually reconnecting. The non-reconnecting component is called the guide field B_g. The guide field is defined by the shear angle θ_{shear}, i.e. the angle between the magnetosheath and magnetospheric magnetic fields

$$\tan\left(\frac{\theta_{shear}}{2}\right) = \frac{B_{0,sh}}{B_g} \tag{3.1}$$

where $B_{0,sh}$ is the reconnecting magnetic field component on the magnetosheath side. In the limit of a shear angle of 180°, the reconnection is anti-parallel, meaning that $B_g = 0$. Observations show that magnetopause reconnection takes place with a large range of shear angles (Trenchi et al. 2008, and references therein).

The characteristics of magnetic reconnection at the magnetopause are deeply affected by the asymmetries across the boundary. The main asymmetries concern the magnetic field and the density across the boundary so that typical values are $B_{sh}/B_{sp} \sim 0.5$ and $n_{sh}/n_{sp} \sim 10$ where the subscript sh and sp indicates the magnetosheath and the magnetosphere, respectively. Typical values of the main physical quantities are summarized in the tables of Fig. 3.2 for both the magnetospheric and magnetosheath side of the magnetopause boundary. The tables have been adapted from Cassak and Fuselier (2016). The conditions on the magnetospheric side are quite stable while the parameters on the magnetosheath side are determined by the solar wind conditions that can change rapidly.

The asymmetry is responsible for the modification of the standard 2D picture of reconnection that is presented in Sect. 2.6 in the following ways

1. the inflows are asymmetric. In particular, the inflow region is more narrow on the magnetospheric side where the magnetic field is larger (Cassak and Shay 2007);
2. the X-point and the stagnation point are not co-located most of the time. In particular, the X-point moves towards the magnetosheath while the stagnation point moves towards the region with the smaller mass flux into the diffusion region. Since the mass flux can be written in terms of n/B the displacement of the stagnation point depends on both magnetic field and density asymmetry. The difference

Magnetospheric parameter	Typical value
n_{ms} (cm^{-3})	0.1
B_{ms} (nT)	56
$T_{i,ms}$ ($\times 10^5$ K)	2.4×10^3
$c_{A,ms}$ (km/s)	3.9×10^3
β_{ms}	0.27

Subsolar magnetosheath parameter (Derived)	Mean	Most probable	Median	5–95 % range start	5–95 % range stop
n_{sh} (cm^{-3})	34.8	20	27.6	12	80
B_{sh} (nT)	24.8	20.4	22.4	8.8	39.6
$T_{i,sh}$ ($\times 10^5$ K)	12	5	9.5	1	30
$c_{A,sh}$ (km/s)	92	99	93	55 (21)[a]	97 (250)[a]
β_{sh}	2.4	0.8	1.8	0.5	5.3

Fig. 3.2 Typical plasma parameters at the magnetospheric side (top) and at the magnetosheath side (bottom) of the Earth's subsolar magnetopause. Density n, magnetic field strength B, ion temperature T_i, ion Alfvén speed c_A, and plasma beta β are reported. The parameters on the magnetosheath side are computed from observations in the solar wind with the assumptions that across the bow shock the density and magnetic field strength increase by a factor of 4, the solar wind ion temperature increases by a factor of 10 across the bow shock and that there is no further change in the shocked solar wind plasma as it convects from downstream of the bow shock to the subsolar magnetopause. [a]The Alfvén speeds in parenthesis are derived assuming anti-correlation between the solar wind density and interplanetary magnetic field strength. Adapted by permission from Springer Nature: Springer, Magnetic Reconnection. Concepts and Applications. edited by W. Gonzalez and E. Parker. © 2016 (Cassak and Fuselier 2016)

in location of the X-point and the stagnation point implies that there is a net bulk flow of plasma across the X-line during asymmetric reconnection (Cassak and Shay 2007);
3. the quadrupolar pattern of the Hall magnetic field is distorted. In particular, the extension of the lobe is reduced on the magnetospheric side. Depending on the intensity of the asymmetry, the Hall magnetic field can degenerate in a almost bipolar pattern. Correspondingly, the pattern of the Hall currents is also modified and the Hall electric field is also asymmetric, being stronger on the magnetospheric side of magnetopause. The change of sign of the Hall electric field is displaced towards the magnetosphere;
4. The peak of the out-of-plane current density together with the ion and electron jet reversal are biased towards the magnetosphere;
5. the gradients (e.g. the temperature gradient) across the boundary are displaced towards the magnetosphere (Egedal et al. 2011).

Figures 3.3 and 3.4 give a typical example of asymmetric reconnection. In particular, Fig. 3.3 shows the evolution of several quantities during a magnetopause crossing during a reconnection event reported by Lavraud et al. (2016). Before discussing in detail, note that the physical quantities in both Figs. 3.3 and 3.4 are presented in the local magnetopause coordinate system LMN. Details about this coordinate system can be found in Appendix A. Qualitatively, the reconnecting magnetic field is

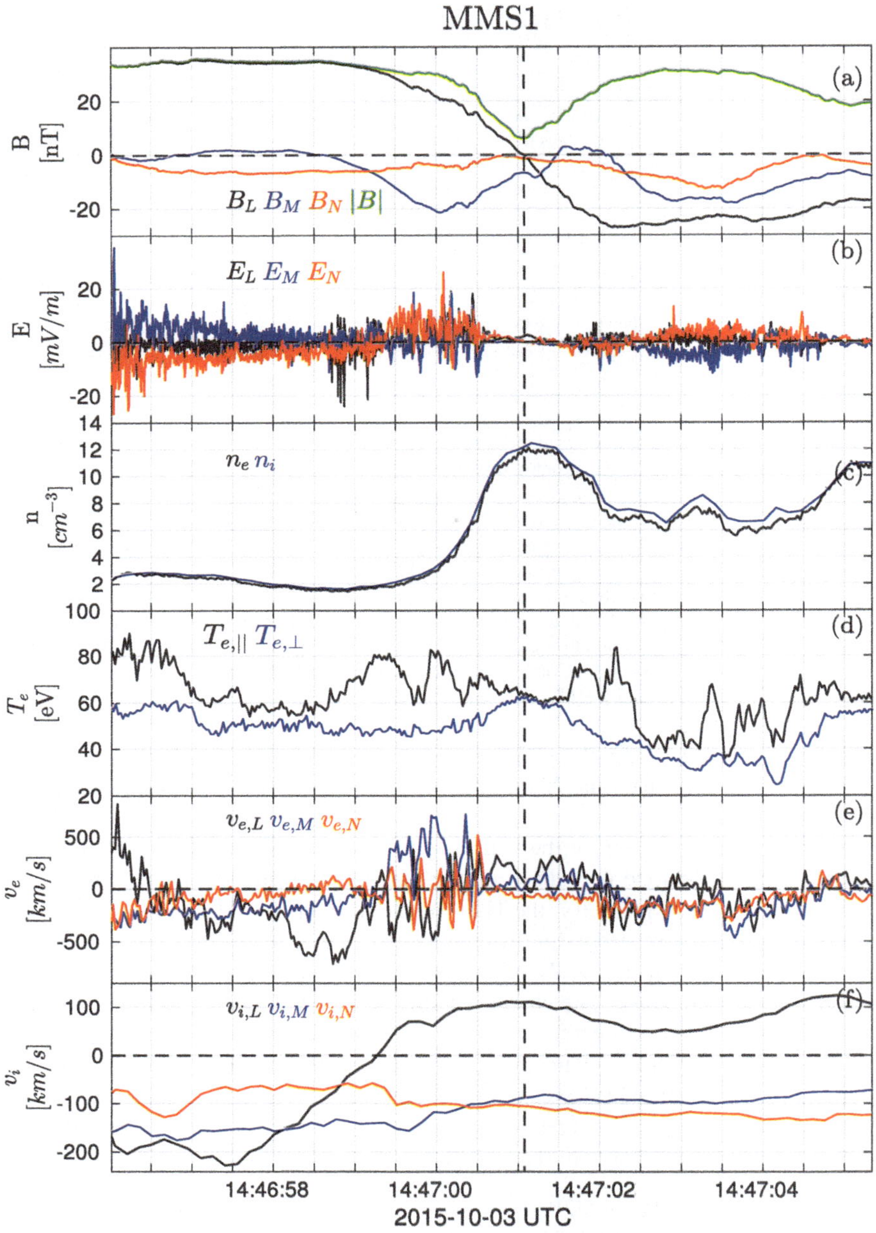

Fig. 3.3 **a** Three components of the magnetic field and the magnetic field strength; **b** three components of the electric field; **c** electron and ion number density; **d** parallel and perpendicular (to the magnetic field) electron temperature; **e** three components of the electron velocity; **f** three components of the ion velocity. All data are shown in the LMN coordinate system obtained as in Lavraud et al. (2016). The vertical dashed line indicates the magnetic field minimum

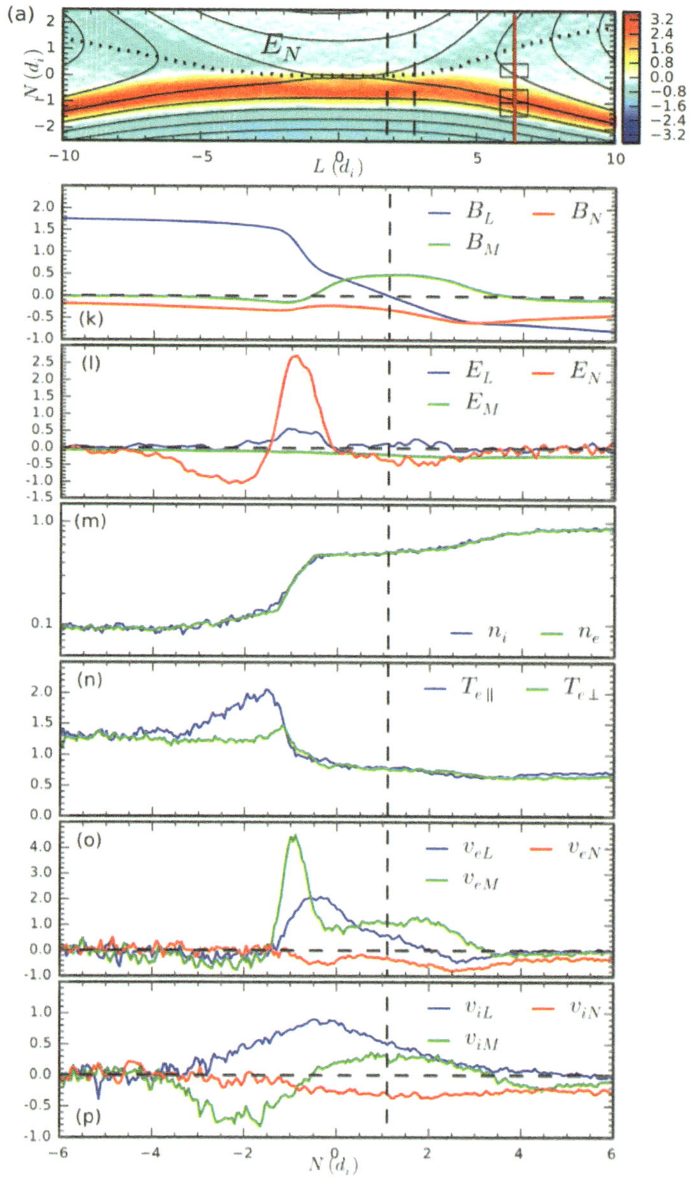

Fig. 3.4 **a** Contour plot of the Hall electric field E_N. The red solid line indicates where the cuts shown in the other panels were taken ($L = 6.35\,d_i$). **k** Three components of the magnetic field; **l** three components of the electric field; **m** ion and electron number density; **n** parallel and perpendicular (to the magnetic field) electron temperature; **o** three components of the electron velocity; **p** three components of the ion velocity. Adapted from Shay et al. (2016). © 2016 by the American Geophysical Union

expected to be directed along L – as well as the ion and electron jets – while the Hall magnetic field is directed along M. The LMN system used in Fig. 3.3 is the same that has been adopted in Lavraud et al. (2016). Figure 3.3a shows the magnetic field. The magnetic field component B_L is changing from positive to negative values, meaning that the spacecraft is crossing the magnetopause going from the magnetosphere side to the magnetosheath side of the boundary. This scenario is also supported by the evolution of the density shown in Fig. 3.3c. The vertical dashed line indicates the magnetic field minimum. The M component of the magnetic field presents a dipolar signature that indicates that MMS is crossing two of the lobes of the quadrupolar Hall field, Fig. 3.3a. However, the dipolar signature is distorted because of the asymmetry. An ion (and electron) jet reversal is also observed and the inversion point is shifted towards the magnetosphere (Fig. 3.3d, e). This is the case also for the peak of $v_{e,M}$, as expected from the previous discussion. The enhancement of E_N (the Hall electric field) is not so clear for this event but one can see that the electric field activity is biased towards the magnetosphere (Fig. 3.3a).

Simulations results provide a picture which in consistent with observations, as shown in Fig. 3.4. Figure 3.4 has been adapted from Shay et al. (2016) showing an almost bipolar Hall magnetic field across the current sheet (panel (k)); the electric field (panel (l)) and velocity (panel (o)–(p)) peaks are located in the magnetospheric side as well as the temperature gradient.

Together with the asymmetry, the intensity of the guide field plays a role in modifying the symmetric picture presented in Sect. 2.6. It has been shown that when both the guide field and asymmetry increase, the symmetric Hall magnetic field pattern gradually degenerates into an asymmetric (bipolar) pattern (Eastwood et al. 2013; Pritchett and Mozer 2009; Tanaka et al. 2010). However, an asymmetry characterized by $n_{sp} \sim 0.5\ n_{sh}$ and $B_{sp} \sim B_{sh}$ combined with a guide field $B_g = 0.4\ B_0$ is not able to completely dismiss the quadrupolar pattern of Hall reconnection (Peng et al. 2017). A similar result is obtained with a larger asymmetry ($n_{sp} \sim 0.1\ n_{sh}$ and $B_{sp} \sim 2\ B_{sh}$) and smaller guide field ($B_g = 0.2\ B_0$) (Chen et al. 2017).

References

P.A. Cassak, S.A. Fuselier, in ed. by W. Gonzalez, E. Parker. *Magnetic Reconnection. Concepts and Applications* (Springer, Cham, 2016), p. 213

P.A. Cassak, M.A. Shay, Phys. Plasmas **14**, 102114 (2007)

L.-J. Chen et al., J. Geophys. Res. **122**, 5235 (2017)

J.W. Dungey, Phys. Rev. Lett. **6**, 47 (1961)

J.P. Eastwood et al., Plasma Phys. Control. Fusion **55**, 124001 (2013)

J. Egedal, A. Le, P.L. Pritchett, W. Daughton, Phys. Plasmas **18**, 102901 (2011)

C.M. Komar, Ph.D. thesis, West Virginia University (2015)

B. Lavraud et al., Geophys. Res. Lett. **43**, 3042 (2016)

F.Z. Peng et al., J. Geophys. Res. **122**, 6349 (2017)

P.L. Pritchett, F.S. Mozer, J. Geophys. Res. **114**, A11210 (2009)

M.A. Shay et al., Geophys. Res. Lett. **43**, 4145 (2016)

K.G. Tanaka, M. Fujimoto, I. Shinohara, Int. J. Geophys. **2010**, 202583 (2010)

L. Trenchi et al., J. Geophys. Res. **113**, A07S10 (2008)

Chapter 4
Methods of Spacecraft Data Analysis

This chapter is divided into two parts. The first section is devoted to the presentation of the Magnetospheric MultiScale (MMS) mission, a NASA mission which has been launched in 2015 and whose data are used throughout the Thesis. Information about the spacecraft, the instrumentation on board, the orbit and data products are provided. Three sections are focused on the description of the instruments that are widely used in this Thesis, namely the FluxGate Magnetometer providing magnetic field measurements, the Spin-Plane Double Probe and the Axial Double Probe providing electric field measurements and the Fast Plasma Investigation, providing electrons and ions distribution functions and their momenta.

The second part of the chapter presents the data analysis techniques that are used for the observational study reported in Chap 7. These methods are both single and multi spacecraft methods and they provide information about the orientation, velocity, dimensionality of the structure sampled by the spacecraft. A method to infer the magnetic field topology is also presented.

4.1 Magnetospheric MultiScale (MMS) Mission

4.1.1 Overview and Objectives

The NASA's Magnetospheric MultiScale (MMS) mission is composed by four identically equipped satellites flying in a tetrahedral formation in near-Earth space. It has been launched in March 2015. The main scientific goal, as stated in Burch et al. (2016b) is to

G. Cozzani, *Microphysics of Magnetic Reconnection in Near-Earth Space*, Springer Theses, https://doi.org/10.1007/978-3-030-56142-0_4

understand the microphysics of magnetic reconnection by determining the kinetic processes occurring in the electron diffusion region that are responsible for collisionless magnetic reconnection, especially how reconnection is initiated.

Therefore, the spacecraft instruments and the tetrahedral configuration have been designed to reach time and spatial resolutions that allow the study of electron-scale physics. MMS is not the first mission to sample the magnetosphere and the adjacent space to be composed by four spacecraft. Indeed, ESA's Cluster mission (Escoubet et al. 1997) (launched in 2000) has also four probes and it has lead to critical advance in the understanding of magnetic reconnection. The presence of four probes in formation is fundamental since it allows to discriminate the temporal from the spatial variations and to compute the spatial gradients, as detailed in Sect. 4.2.2. The novelty of MMS lies specifically in the orbit (equatorial, while Cluster orbit is polar), in the inter-spacecraft separation's range, in the particles data resolutions and in the electric field measurements. Being the electron-scale physics the main target, the minimum inter-spacecraft is less than 10 km which is comparable to the electron inertial length d_e at the magnetopause and smaller than d_e at the magnetotail. The maximum inter-spacecraft separation is of the order of 400 km. Then, a new approach for particle measurements has been devised for MMS in order to achieve unprecedented time resolution for particle data. While previous magnetospheric missions have often used the rotation of the spacecraft to acquire 3D particle distributions, MMS Fast Plasma Investigation instrument (FPI, see Sect. 4.1.9 for further details) is composed by multiple sensors that cover 4π steradians in 30 ms for electron and 150 ms for ions. On Cluster, the particle measurement are acquired every 2 s (electrons) and 4 s (ions). The MMS spacecraft rotation period of MMS is 20 s.

The required time resolutions derive from the fact that the electron and ion inertial lengths are the spatial scales that have to be resolved by MMS. Moreover, reconnection layers and the associated diffusion regions are usually moving. This motion is oscillatory at the dayside magnetopause and it is usually tailward in the magnetotail and typical velocities range from tens of km/s to 100 km/s. An electron diffusion region of width $d_{EDR} \sim 5$ km moving at 50 km/s would be observed by only one of the MMS spacecraft for a time interval of 0.1 s. In order to have at least 3 measures within the EDR a time resolution of 0.03 s is required. Another novelty of MMS as compared to previous missions is the presence of an axial probe that allows for 3D electric field measurements. Further details about the electric field data are provided in Sects. 4.1.6 and 4.1.7.

4.1.2 The Spacecraft

The MMS constellation consists of four identically instrumented spacecraft. Each satellite has an octagonal shape and it is approximately 3.5 m wide and 1.2 m high and at launch, with a full load of propellant, it weighs approximately 1360 kg. The

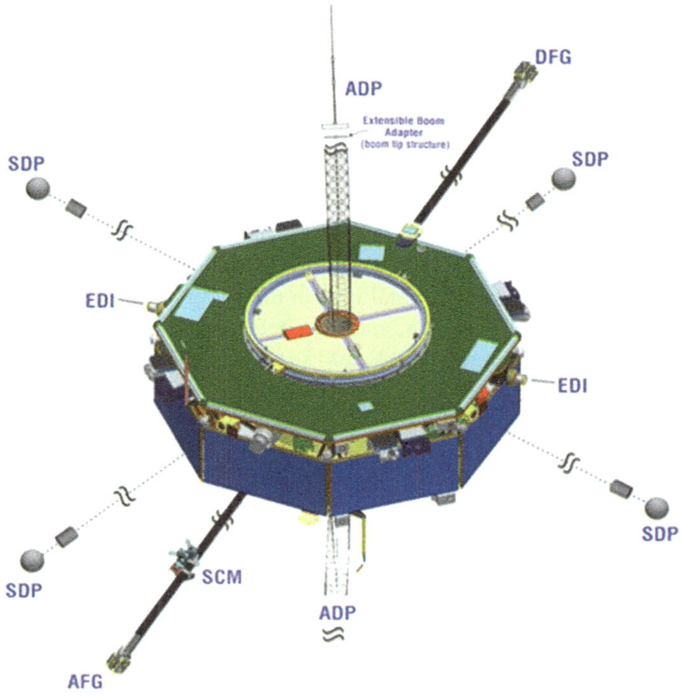

Fig. 4.1 FIELDS sensors on a MMS spacecraft. Adapted from Torbert et al. (2016b)

instruments onboard each spacecraft are listed below. The FIELDS suite (Fig. 4.1) consists of the instruments measuring the magnetic and electric field. In particular,

- the FluxGate Magnetometer (FGM) measures the magnetic field at a sampling frequency of $128\,s^{-1}$. The accuracy of the instrument is $\sim$0.1 nT for the DC field. It is composed by the Analog Flux-Gate (AFG) and the Digital Flux-Gate (DFG);
- the Search Coil Magnetometer (SCM) measures the magnetic field fluctuations in the 1 Hz–6 kHz nominal frequency range;
- the Electric field Double Probe (EDP) is composed by the Spin-plane Double Probes (SDP) and the Axial Double Probes (ADP) and it measures the 3D electric field in the frequency range DC–100 kHz.
- the Electron Drift Instrument (EDI) measures the magnetic and electric field. The cadence is smaller compared to other instruments (10 samples/s).

The instruments measuring particles are:

- the Fast Plasma Investigation (FPI) instrument, composed by four Dual Electron Spectrometers (DES) and four Dual Ion Spectrometers (DIS) per spacecraft. FPI combines the data from DES and DIS providing the velocity-space distribution functions of electrons and ions with a time resolution of 30 ms and 150 ms respectively, in the energy range [10 eV, 30 keV];

- the Hot Plasma Composition Analyzer (HPCA), providing the composition-resolved 3D ion energy distributions of H^+, He^{++}, He^+, and O^+ in the energy range [1 eV, 40 keV];
- the Energetic Particle Detector (EPD), including an Energetic Ion Spectrometer (EIS) and an all-sky particle sampler called the Fly's Eye Energetic Particle Sensor (FEEPS). They measure the energy-angle distribution and composition of ions in the energy range [20 keV, 500 keV] with a sampling frequency of < 30 s, the energy-angle distribution of the total ions in the energy range [45 keV, 500 keV] at a sampling frequency of <10 s, and the energy-angle distribution of energetic electrons in the energy range [25 keV, 500 keV] with a sampling frequency <10 s.

The Active Spacecraft Potential Control (ASPOC) generates beams of indium ions to limit positive spacecraft potentials within +4 V.

Since data from FGM, EDP and FPI instruments are extensively used in this Thesis a detailed description of these instruments is given in Sects. 4.1.6, 4.1.7, 4.1.8 and 4.1.9.

4.1.3 Orbit and Mission Phases

MMS orbit is a low-inclination (28 degrees) elliptical orbit with a perigee of 1.2 R_E (Earth radii). The prime mission (September 2015–September 2017) is composed by two phases during which MMS have sampled the regions in near-Earth space where the probability of encountering the diffusion region is highest i.e. the dayside magnetopause (Phase 1) and the magnetotail (Phase 2). The apogee increases accordingly, in particular it is 12 R_E during Phase 1 and it is increased to reach 25 R_E during Phase 2 (Fig. 4.2). The orbital segment in which the probability to observe the magnetic reconnection process (the magnetopause boundary and the magnetotail neutral sheet) are the so-called Regions of Interest (ROI). At the dayside, the ROI is the portion for which MMS is further than 9 R_E from Earth. In the magnetotail, MMS has a high probability to cross the diffusion region when the distance from Earth ranges from 15 R_E and 25 R_E. In addition, the nominal region of interest is allowed to shift almost weekly according to expected magnetospheric conditions in order to optimize science data collection. As it will be detailed in the following the highest resolution data are acquired only within the ROI. The portions of the orbit that are not considered of scientific interest are used for calibration, health/safety, and maintenance activities.

Phase 2 officially ended on September,1 2017. During the following Phase 3, MMS orbit has maintained the same apogee and perigee but is has started sampling the dayside again. For this reason, MMS is also currently acquiring data in the solar wind. To provide better observations in the solar wind, the apogee has been increased to 29 R_E during Phase 4 of the extended mission. The mission phases and their duration are summarized in Table 4.1.

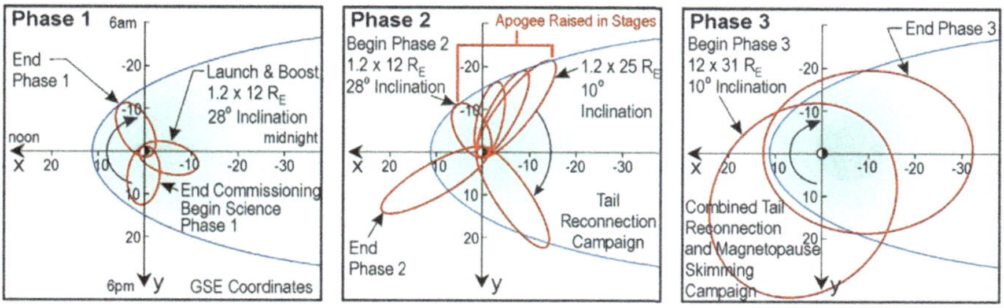

Fig. 4.2 MMS orbit during (left) Phase 0 and 1, (center) the transition from Phase 1 (apogee of 12 R_E) to Phase 2 (apogee of 25 R_E), (right) Phase 3. The orbits are shown in the x–y plane (GSE coordinates). (Credits: NASA/SwRI)

Table 4.1 MMS mission phases during the prime and extended mission

MMS mission phases		
Phase	Period of time	Apogee (R_E)
Phase 0 (Commissioning)	15/03/2015–30/08/2015	12
Phase 1 (Prime)	01/09/2015–30/01/2017	12
Phase 2 (Prime)	31/01/2017–01/09/2017	12 → 25
Phase 3 (Extended)	02/09/2017–27/09/2018	25
Phase 4 (Extended)	08/09/2018–Present	25 → 29

4.1.4 The Constellation

Throughout the regions of interest the four MMS spacecraft are kept in a tetrahedral configuration. The inter-spacecraft separation changes depending on the region to investigate. A lower range (10–160 km) will be used on the dayside ($d_e \sim 1$ km) while a higher range (30–400 km) will be used on the night side ($d_e \sim 30$ km). The optimal configuration is a regular tetrahedron with separation appropriate to the region of interest crossed by the spacecraft. A quality factor Q_C has been already defined for the Cluster mission $Q_C = V_{act}/V_{reg}$, where V_{act} is the volume of the actual tetrahedron while V_{reg} is the volume of the regular tetrahedron with the side equal to the average of the inter-spacecraft separations of the actual constellation. Since the spacecraft separation in function of the position along the orbit are also crucial for MMS measurements, the new quality factor Q_{MMS} is defined as $Q_{MMS} = Q_C \cdot Q_s$ where Q_s takes into account the inter-spacecraft separation. Details about Q_s definition and computation are given in Fuselier et al. (2016).

4.1.5 Mission Operations and Data Products

There are two basic instrument science operation modes (slow survey and fast survey) and three data acquisition rates: slow, fast and burst (Baker et al. 2016). Within the

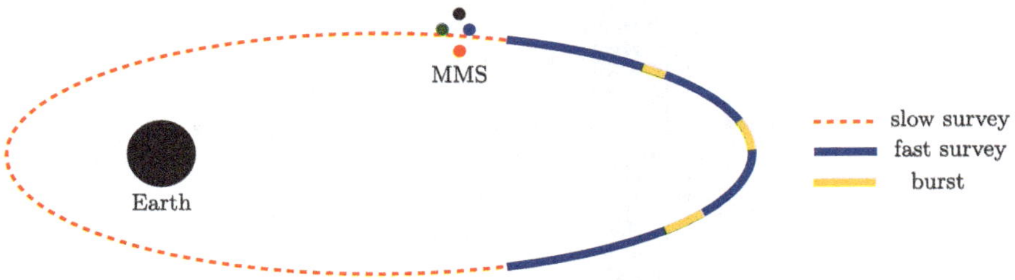

Fig. 4.3 Ecliptic-plane scheme of MMS orbit (red dashed). The ROI is the blue orbital segment while the burst segments are in yellow

regions of interest the fast survey mode is used (Fig. 4.3). The data are acquired in both the fast and burst rates and they are stored on board at the highest resolution. The time resolution of the fast mode data is comparable to that of previous magnetospheric mission. The data acquired at fast rates are all transmitted to the ground. The same treatment is not possible for the burst data since their volume exceeds the capability of the MMS telemetry transmission system.

Even though almost 75% of the telemetry bandwidth is allocated to burst data, this allows the transmission of only ~2% of the totality of the burst data. However, it is possible to store on board several orbits of data and a portion of data with the highest scientific interest is transmitted daily. But how are the most scientifically valuable data selected? The data selection process is partly automated. While MMS is in the ROI, each instrument automatically scans its measurements and assigns quality factors to each 10 s segment. High quality factors are given to data that show strong magnetic field variations, large electric field and electron current density, the presence of ion and electron beams. All these signatures may indicate that magnetic reconnection is ongoing. Then, the quality factors of the different instruments are combined to provide a Figure Of Merit (FOM) for the data segment.

The selection is then checked by a Scientist-in-the-Loop (SITL) that examines the automatically assigned FOM. The SITL can validate the selection, add or delete data segments or change their downlink priorities. Furthermore, the SITL fills in a report in which each selection is commented with the physical phenomenon observed in each selected time segment. The SITL operates following the FOM guidelines which depends on the mission phase. For example, during Phase 1 (dayside reconnection being the main scientific goal) the highest FOM (150–199) is supposed to be assigned to complete high-shear magnetopause crossings (Category 1 data). Partial magnetopause crossings usually have a lower FOM (100–129, Category 2) but if diffusion region's signatures are present they are considered Category 1. Magnetosheath current sheets with reconnection signatures and Flux Transfer Events (FTEs) with jets signatures are Category 1 data as well.

Multiple categories of data are produced and they are characterised by different levels of refinement and calibration. The *level 1* data are raw data at full resolution; the *QuickLook* data are scientific data products that are rapidly generated using simplified processing algorithms. These data provide basic scientific insight but calibrations are still provisional at this stage. Their generation occurs after some hours from the reception. *Level 2* are the calibrated version of *level 1* data, they have the same time resolution but *Level 2* data have higher scientific value. *Level 3* or mission-level data are produced on an event basis and they have the highest sophistication in terms of calibration and treatment.

Throughout this Thesis, *level 2 burst* data will be used. The MMS mission has an open data system with no proprietary data periods. All data can be found in the MMS Science Data Center (SDC) (https://lasp.colorado.edu/mms/sdc/public/).

4.1.6 Magnetic Field Measurement: The FluxGate Magnetometer

The magnetic field measurements are acquired using two triaxial fluxgate magnetometers on each spacecraft. Each magnetometer is mounted on the end of a boom 5 m long. The difference among the two instruments lies in the electronic unit to which they are connected. One is connected to the Digital FluxGate (DFG) provided by the Space Research Institute of the Austrian Academy of Sciences, the other to the Analog FluxGate (AFG), provided by the University of California, Los Angeles. The technical details about DFG and AFG can be found in Russell et al. (2016) and lie beyond the scope of this Thesis. The FluxGate Magnetometer (FGM) measures the magnetic field at a sampling frequency of $16\,s^{-1}$ in *survey* mode and $128\,s^{-1}$ in *burst* mode. The accuracy of the instrument is $\sim$0.1 nT for the DC field.

Fluxgate sensors are ferromagnetic cores around which are wrapped two coil windings: the drive winding and the sense winding. When an alternating current is present in the drive winding an induced current is present in the sense winding. If there is no external magnetic field, the two currents cancel out. In presence of an external magnetic field, an offset between the drive current and the sense current is produced yielding to a net magnetic flux in the sense winding that can be measured as a voltage using Faraday's law.

4.1.7 Electric Field Measurements: The Spin-Plane Double Probe

The Spin-plane Double Probe (SDP) instrument (Lindqvist et al. 2016) measures the electric field in the spacecraft spin plane. Together with the ADP instrument (see Sect. 4.1.8), SDP measures the 3-D electric field over the frequency range DC–100 kHz. The SDP instrument is the result of the cooperative effort among the University of New Hampshire, the Royal Institute of Technology (KTH), Swedish Institute of Space Physics (IRFU), the University of Oulu, and the Laboratory for Atmospheric and Space Physics (LASP) of the University of Colorado. The electric field along two

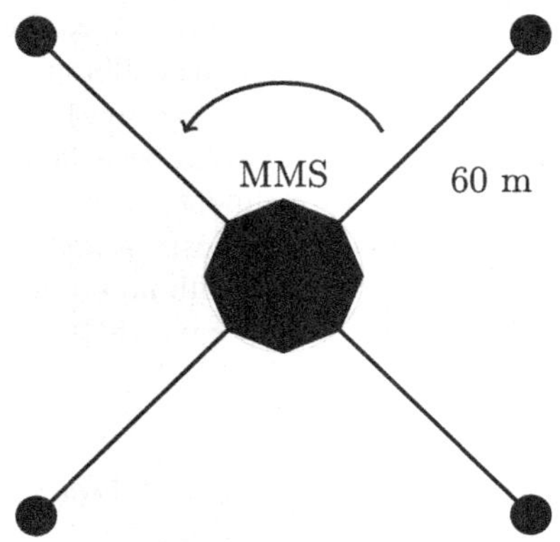

Fig. 4.4 SDP booms and probes in the x–y plane (spacecraft coordinates)

perpendicular directions is computed from the difference of potential between two spherical titanium-nitride electrodes, each of diameter 8 cm at the end of wire booms with a length of 60 m, as represented in Fig. 4.4. The potential difference between two opposed probes yields a measurement of the electric field in the direction along the axis defined by the two probes and the boom. Since two pairs of probes are present, two orthogonal components of the electric field in the plane of the spacecraft spin are measured. In general, a larger distance between the probes results in a larger potential difference, which is easier to measure. Furthermore, a longer boom allows for a smaller error in the electric field measurement since it permits to partially overcome the effects of the Debye shielding as well as the effects of the photoelectrons.

One of the main issues concerning electric field measurements is that a conductive sphere into a sunlit plasma does not sit at the same potential as the ambient plasma. This is due to the interaction among the probe, the photons coming from the Sun that cause photoelectron emission and the ions and electrons which compose the plasma. The production of photoelectrons leads the sphere to be charged positively while the colliding particles determine a negative or positive charge. These effects can be summarized as

$$I_e + I_i + I_{ph} = 0 \tag{4.1}$$

where I_e is the current of electrons coming to the probe ($I_e < 0$), I_i is the current of ion coming to the probe ($I_i > 0$) and I_{ph} is the current of photoelectrons escaping from the probe ($I_{ph} > 0$). Usually this reduces to $I_e + I_{ph} = 0$ since the ion contribution is negligible, especially in the sunlight and in the low-density magnetosphere. The relation $I_e + I_{ph} = 0$ provides the probe potential which can be many volts positive in a low-density plasma.

In order to have a probe potential which is closer to zero, a negative bias current I_b is added. In this way, the potential measurement is less sensitive to fluctuations in the plasma electron current. The nominal value range of the bias current in the sunlight is $[-200, -100]$ nA while I_b is set to zero during eclipse. The value of the bias current is mainly determined by the photocurrent, which depends on the solar photons flux as well as on the properties of emission of the probe surfaces. For this reason, the bias current is changed along the mission depending on the solar photons flux.

The SDP instrument runs in several modes. The sampling frequencies are 8 Hz in *slow survey* mode, 32 Hz in *fast survey* mode, and 1024, 8192, or 65536 Hz in *burst* mode. After reception on the ground, SDP data are combined with ADP data to provide the full electric field vector.

4.1.8 Electric Field Measurements: The Axial Double Probe

The Axial Double Probe (ADP) instrument (Ergun et al. 2016) measures the electric field along the spin axis of the spacecraft. Together with the SDP instrument (see Sect. 4.1.7), ADP measures the 3-D electric field over the frequency range DC–100 kHz. ADP has been developed in collaboration among the Laboratory for Atmospheric and Space Physics (LASP) of the University of Colorado, the University of New Hampshire and the Royal Institute of Technology (KTH). The presence of an axial probe allowing for the measurement of the complete 3D electric field vector is a novel feature of MMS compared to Cluster, where the axial component is computed using the relation $\mathbf{E} \cdot \mathbf{B} = 0$. The physical background behind the ADP instrument operation is analogous to SDP, the electric field is computed from the difference of the potential between two probes. In this case, the two probes are positioned on top of antennas directed along the z axis of the spacecraft, namely the spin axis. The antenna lengths are constrained by mechanical limitations, which include deployment of stiff booms while preserving spacecraft stability. For this reason, the ADP antennas are smaller than the SDP ones. Their length is 30 m tip-to-tip. The ADP booms and probes are designed to have cylindrical symmetry, primarily to eliminate modulation of the photoelectron current as the spacecraft rotates. The design also optimizes symmetry between the +z and -z sensors so that opposing sensors experience nearly the same potential and they have nearly identical photoelectron environments.

The sampling frequencies are the same of SDP.

4.1.9 Particle Measurements: The Fast Plasma Investigation

The Fast Plasma Investigation (FPI) instrument (Pollock et al. 2016) provides 3D phase space distributions of electrons and ions in the energy range [10 eV, 30 keV] with the unprecedented time resolutions of 30 ms and 150 ms for electrons and ions respectively. The FPI was collaboratively developed by institutions in Japan, France and USA with Goddard Space Flight Center (GSFC) as the Lead Co-Investigator institution.

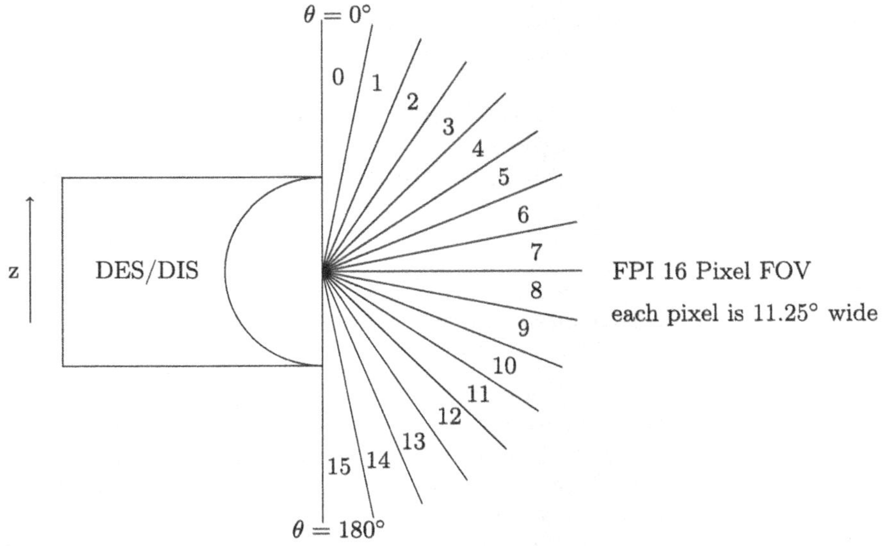

Fig. 4.5 Schematic description of the Field Of View (FOV) of the *top hat* electrostatic analyzers (ESA) used in the FPI instrument

The FPI consists of eight electron spectrometers and eight ion spectrometers mounted on each spacecraft. These are organised in pairs of Dual Electron Spectrometer (DES) and Dual Ion Spectrometer (DIS) in back-to-back configuration around the spacecraft as shown in Fig. 4.6. This configuration allows to reach a time resolution that does not depend on the spacecraft spin, as has been common for magnetospheric missions (e.g. Cluster).

The DES and DIS composing the FPI instruments are *top hat* electrostatic analyzers (ESA). The *top hat* ESA geometry has been extensively used in the past and it has been presented by Carlson et al. (2001) within the description of the ion and electron plasma experiment on the FAST (Fast Auroral Snapshot) satellite. The *top hat* ESA is able to select and count the particles with a specific velocity direction and an energy/charge ϵ within the interval $[\epsilon^* - \Delta\epsilon, \epsilon^* + \Delta\epsilon]$. It usually allows simultaneous observations over an angular range of 360° but FPI only uses half of the *top hat* Field Of View (FOV), namely 180°. The FOV is divided in 16 pixels with a width of 11.25° each (Fig. 4.5). We refer to the angular direction describing the FOV as the polar angle θ, it corresponds to the angle opening from the spacecraft z-axis. Another angular direction – used in the following – is ϕ, the azimuth angle which is defined in the spacecraft plane x–y and opens from the spacecraft +x axis with a positive right hand rotation about the spacecraft +z axis ($0° < \phi < 360°$).

For each DES (and DIS) the two undeflected FOV are separated by 45° along ϕ (see Fig. 4.6). Then, in order to reach a full sky view, a FOV deflection with a maximum angle of $\pm 16.875°$ along ϕ is implemented. This is accomplished by applying positive voltage to curved electrodes located just inside of the sensor entrance apertures. The

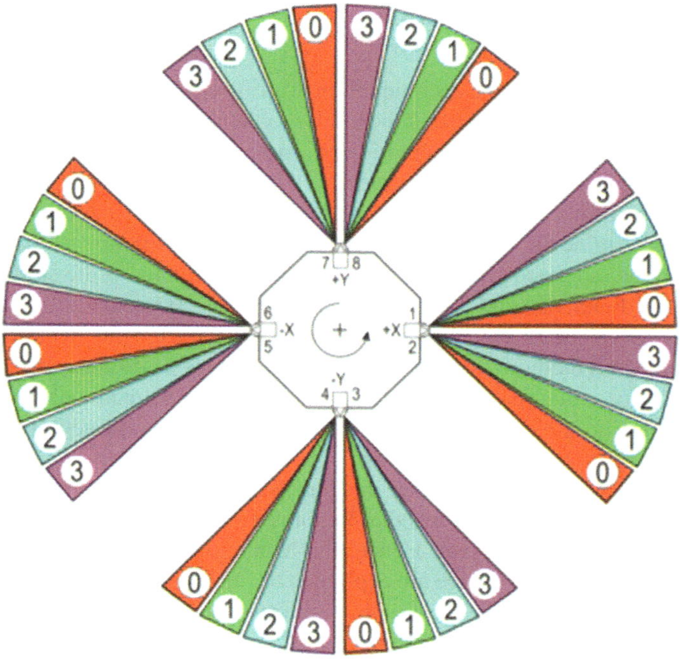

Fig. 4.6 Azimuth coverage of the eight DES (or DIS). The DES (or DIS) locations are indicated with the numbers from 1 to 8 on the spacecraft. Each spectrometer exercises four deflected fields of view with a maximum deflection of ~17°. Adapted from Pollock et al. (2016)

azimuth coverage is shown in Fig. 4.6. More details about the design and functioning of FPI and DES/DIS can be found in Pollock et al. (2016).

The particles that are passed by the ESA are then detected with Micro Channel Plates (MCP) which amplify each incoming electron or ion into a pulse of outgoing electrons. The number of particles per pulse is characterized with a histogram. The fundamental data products are the so-called *skymaps* that contain the raw counts from DES and DIS. These are count rate arrays ($32\epsilon \times 32\phi \times 16\theta$) accumulated every 30 ms for DES and 150 ms for DIS. The final product is a distribution function in the coordinates ($\epsilon, \phi, \theta, t$) for ions and electrons. Finally, the distribution function is properly integrated in order to obtain the momenta.

The fact that the sensors are distributed all around the spacecraft allows to have a full azimuthal sampling without depending on the spin of the spacecraft leading to the unprecedented time resolution of the FPI instrument. On the other hand, an unavoidable consequence of this approach is that the 3D phase space density for either species must be gathered together from the measurements of eight different spectrometers. This represents a complex challenge in terms of inter-calibration that has been handled both during the development of the instrument and during the mission phases (Pollock et al. 2016).

4.2 Data Analysis Techniques

This section summarizes the techniques that are used to perform the MMS data analysis reported in Chap. 7. One of the main problems to overcome when analysing spacecraft measurements is the definition of a proper coordinate frame. Indeed, it is desirable to analysed the examined structures – such as current sheets or flux ropes – in their proper co-moving frame. There are several methods that have been developed in order to determine the coordinate system relative to the structure to analyse. Other information that are needed are the estimation of the velocity of the structure, of its characteristic dimension and dimensionality (1D, 2D or 3D structure). The methods providing such information can be single or multi spacecraft. In this section I describe a single spacecraft method (the Minimum Variance Analysis (Sonnerup and Scheible 1998)) and three multi-spacecraft methods, the Minimum Directional Derivative (or Difference) analysis (Shi et al. 2005), the timing analysis (Harvey 1998) and the curlometer method (Robert et al. 1998).

Since the data analysis presented in Chap. 7 is focused on the diffusion region at the magnetopause, the structure that is going to be crossed by the spacecraft is the magnetopause current sheet. Hence, the Minimum Variance Analysis and the Minimum Directional Derivative are applied to the magnetic field in order to provide the direction normal to the boundary; the timing analysis is applied to the magnetic field in order to estimate the direction normal to the boundary and the boundary velocity along that direction.

After that, the computation of spatial gradients by means of measurement in four points in space are discussed. The determination of all three components of a spatial gradient requires at least four spacecraft in a tetrahedral constellation. In particular, the curlometer method is presented. The curlometer is a multi-spacecraft technique which provides an estimation of the current density averaged over the spacecraft tetrahedron. It represents a fundamental tool for the current density estimation when the particle data have a much smaller time resolution compared to the magnetic field data (e.g. for Cluster mission). With MMS data, the current density can be computed using the particle data directly. However, the comparison between the current density computed with the curlometer and the particles current density is still useful to identify small scales structures. The curlometer is designed to give an estimation of the curl (specifically of the magnetic field) but it provides as well the gradient of the considered quantity.

The last part of the section is devoted to the description of two methods providing information about the magnetic field topology around the spacecraft tetrahedron: the Poincaré method (Greene 1992) and the First-Order Taylor Expansion (FOTE) method (Fu et al. 2015). We will focus in particular on the latter one. These two methods allow to identify the position of a magnetic null, a location where the magnetic field magnitude vanishes. Since many scenarios of magnetic reconnection are attributed to the magnetic nulls, knowing the magnetic field topology in proximity of the reconnection site is useful to investigate the structure of the diffusion region and to understand the particle dynamics.

4.2.1 Single Spacecraft Methods

Minimum Variance Analysis

The Minimum Variance Analysis (MVA) allows to determine the normal direction to a transition layer (in our case, a current sheet) in the hypothesis of a one dimensional, stationary layer which crosses the spacecraft. In principle, magnetic field time series from a single spacecraft is sufficient to obtain the normal direction to the layer **n**. On the other hand, if more than one spacecraft measure the transition across the layer we can check the consistency of the results issued from the different probes. In addition, we can have information about the stationarity of the layer. The main assumptions imposed by this method are the one-dimensionality and the stationarity of the layer during the crossing. If these hypothesis were strictly fulfilled, the magnetic field along **n** would be constant. Since this is hardly obtained in observations, **n** is computed by minimizing the following expression

$$\sigma^2 = \frac{1}{M}\sum_{m=1}^{M} |(\mathbf{B}^m - \langle \mathbf{B} \rangle) \cdot \mathbf{n}|^2 \tag{4.2}$$

where $\mathbf{B}^m$ is the mth element of the magnetic field time series composed of M measurements while $\langle \mathbf{B} \rangle$ is the average value. The minimization of σ^2 is done with the constraint $|\mathbf{n}^2| = 1$,

$$\frac{\partial}{\partial j}(\sigma^2 - \lambda(\mathbf{n}^2 - 1)) = 0 \tag{4.3}$$

where $j = x, y, z$ and λ is the Lagrangian multiplier. After calculations, the problem reduces to an eigenvalue problem

$$\mathcal{M}\mathbf{n} = \lambda \mathbf{n} \quad \text{where} \quad \mathcal{M}_{\rm ij} = \langle \mathrm{B_i B_j} \rangle - \langle \mathrm{B_i} \rangle \langle \mathrm{B_j} \rangle \tag{4.4}$$

where $i, j = x, y, z$.

To solve the problem, we have to find the eigenvalues $\lambda_1, \lambda_2, \lambda_3$ and the corresponding eigenvectors $\mathbf{v}_1, \mathbf{v}_2, \mathbf{v}_3$ of the matrix $\mathcal{M}$. Since $\mathcal{M}$ is symmetric all the eigenvalues are real. In the following, the relation $\lambda_3 < \lambda_2 < \lambda_1$ holds, so that λ_2 is always the intermediate eigenvalue.

The eigenvector $\mathbf{v}_3$, corresponding to the smallest eigenvalue λ_3, is used as the estimator for the normal to the layer **n** since λ_3 itself represents the variance of the magnetic field along the normal direction (minimum variance). Indeed, the diagonal elements of $\mathcal{M}$ are $\mathcal{M}_{ii} = \langle B_i^2 \rangle - \langle B_i \rangle^2$ which is the definition of the variance of the component $i = x, y, z$ of the magnetic field. All eigenvalues are therefore non-negative as expected. Accordingly, the eigenvectors $\mathbf{v}_1, \mathbf{v}_2$ correspond to maximum and intermediate variance and they are tangential to the transition layer. The orthogonal set ($\mathbf{v}_1, \mathbf{v}_2, \mathbf{v}_3$) provides a suitable set for the local coordinate system describing the layer (e.g. the current sheet).

If $\lambda_3 \ll \lambda_2 \ll \lambda_1$ we are able to easily identify a direction of maximal, intermediate and minimal variance and the matrix $\mathcal{M}$ is said to be non-degenerate. The matrix is said to be degenerated when two of the eigenvalues are of the same order. If $\lambda_3 \ll \lambda_2 \simeq \lambda_1$ it is still possible to identify the minimum variance direction while if $\lambda_3 \simeq \lambda_2 \ll \lambda_1$ we can only infer that the eigenvector corresponding to λ_1 is tangential to the current sheet. Lastly, if $\lambda_3 \simeq \lambda_2 \simeq \lambda_1$ we have no information about the orientation of the current sheet. Empirically, for a relatively small set of data ($M < 50$), the normal direction is considered to be well defined for $\frac{\lambda_2}{\lambda_3} > 10$.

Nested Minimum Variance Analysis

The Nested Minimum Variance Analysis allows to check the result obtained with MVA by validating the stationarity of the normal direction. In addition, this method allows to individuate the optimal time interval to perform the MVA. During a nested analysis, a set of concentric segments of data are centered in the event (e.g. at the center of a current sheet). The first interval that we consider is the shortest one and it idealistically contains only three measurement points: one on one side of the layer, one on the opposite side and one point in the middle. The MVA is applied to this interval. Then, in the following steps, the time segment is enlarged by adding a measure on each side, and the MVA is applied again. For an ideal current sheet, at each step the normal direction is the same and the process could in principle continue indefinitely and it will always provide the same result. In reality, a group of the shortest and longest segments often gives results that are significantly different from those obtained for segments of intermediate length. For the shortest segments, these differences can be caused by low scale structures that the current layer may present. For the longest segments, this is normally due to the presence of magnetic structures, different from the event we are focusing on, that begin to be included in the considered time interval.

Within the intermediate range, the normal direction should be the same, or nearly the same, regardless of segment duration. If this is not the case, the layer stationarity hypothesis is probably not satisfied.

Solar Wind – Magnetosphere Data Propagation

Solar wind measurements are usually taken at large distances away from Earth, for example by the Advanced Composition Explorer (ACE) spacecraft or the WIND spacecraft. In order to be representative for the Earth's upstream magnetopause, the ACE or WIND measurements have to be time shifted. Several methods have been developed to provide an estimation of the propagation delay (Mailyan et al. 2008). The simplest model is the so-called *flat delay* model. It assumes that structures in the solar wind moves with constant convective motion along the Sun-Earth line (x-axis of the GSE coordinate system). The estimated time delay is

$$\tau = \frac{\Delta x}{V_{SW,x}}$$

where Δ_x is the distance between the starting point in the solar wind and the target point in the upstream magnetopause and $V_{SW,x}$ is the x component of the solar wind velocity. Indeed, the solar wind velocity is usually mainly aligned with the Sun-Earth line $V_{SW} \approx V_{SW,x}$.

As a reference, the typical time shift to the Earth from ACE is of the order of one hour, depending on the solar wind speed. More sophisticated methods have been developed (Mailyan et al. 2008, and references therein) but the flat delay model is adopted in Chap. 7 since (i) the only need is to establish whether the IMF is directed southward or northward during the interval of interest for the Electron Diffusion Region analysis; (ii) it has been shown that statistically there is little difference among delays computed with different methods (Mailyan et al. 2008; Case and Wild 2012). Indeed, the flat propagation model provides a delay estimation in agreement with methods of higher sophistication. However, it suffers from larger uncertainties.

4.2.2 *Multi-spacecraft Methods*

The presence of a constellation of spacecraft can provide a consistency check for single spacecraft results and gives also the possibility to have a deeper insight in the spacecraft data.

As for single spacecraft techniques, also in the case of multi-spacecraft analysis some general assumptions are made. Some of them as been already pointed out in the previous Sections but they are summarized here for clarity. The first assumptions are the *planarity* and the *stationarity* of the boundary that crosses the spacecraft tetrahedron. Stationarity can be verified by means of a nested analysis (see Sect. 4.2.1) or using the timing method. As a third assumption, the boundary is generally consider as moving with *constant velocity*. Another important hypothesis is about the *characteristic scale*: it is assumed that the characteristic scale of the observed structure is larger than the spacecraft inter-separation.

In some cases, when applying MVA on different spacecraft, non-collinear normal directions are obtained. This usually results from the combined effect of non-constant motion and curvature of the layer.

Spatial Gradients Estimation

In order to compute the spatial gradients of a quantity $\mathbf{B}(\mathbf{r})$ we use the linear interpolation of $\mathbf{B}(\mathbf{r})$ between the discrete values $\mathbf{B}_\alpha$, $\alpha = 1, \ldots, 4$ which are measured at each vertex of the tetrahedron, located at $\mathbf{r}_\alpha$. A linear interpolation of the quantity

$\mathbf{B}(\mathbf{r})$, $\mathcal{L}(\mathbf{B}(\mathbf{r}))$ can be written as

$$\mathcal{L}(\mathbf{B}(r)) = \sum_{\alpha=1}^{4} \mathbf{B}_\alpha \mu_\alpha(\mathbf{r}) \tag{4.5}$$

where $\mu_\alpha = \nu_\alpha + \mathbf{k}_\alpha \cdot \mathbf{r}$. Under the constraint $\mu_\alpha(\mathbf{r}_\beta) = \delta_{\alpha\beta}$ we obtain $\mu_\alpha = 1 + \mathbf{k}_\alpha \cdot (\mathbf{r} - \mathbf{r}_\alpha)$ and

$$\mathbf{k}_\alpha \cdot (\mathbf{r}_\beta - \mathbf{r}_\gamma) = \delta_{\alpha\beta} - \delta_{\alpha\gamma}. \tag{4.6}$$

This implies that the direction of $\mathbf{k}_\alpha$ is normal to the face of the tetrahedron opposite to its corresponding vertex S_α. Hence, we can write $\mathbf{k}_\alpha$ as

$$\mathbf{k}_\alpha = \frac{\mathbf{r}_{\beta\gamma} \times \mathbf{r}_{\beta\delta}}{\mathbf{r}_{\beta\alpha} \cdot (\mathbf{r}_{\beta\gamma} \times \mathbf{r}_{\beta\delta})} \tag{4.7}$$

where $\alpha, \beta, \gamma, \delta = 1 \ldots 4$ and $\mathbf{r}_{\alpha\beta} = \mathbf{r}_\alpha - \mathbf{r}_\beta$. The other vectors are obtained through cyclic permutation of the indices. The vectors $\mathbf{k}_\alpha$ are the reciprocal vectors. They are proportional to the area of the face of the tetrahedron opposite to their corresponding vertex S_α and inversely proportional to the volume of the tetrahedron.

Now, if $\mathcal{G}[\mathbf{B}](\mathbf{r})$ is the tensor of rank two describing the gradient of a vector field $\mathbf{B}(\mathbf{r})$ and $\mathbf{G}[u](\mathbf{r})$ is the vector describing the gradient of a scalar $u(\mathbf{r})$, we have

$$\mathbf{k}_\alpha = \mathbf{G}[\mu_\alpha] \tag{4.8}$$

In order to compute the gradient of a linearized vector field $\mathbf{B}(\mathbf{r})$ we combine Eqs. 4.5 and 4.8 and we use that the linear interpolation operator $\mathcal{L}$ is a linear operator $(\mathbf{G}[\mathcal{L}(u)] = \mathcal{L}(\mathbf{G}[u])$ and $\mathcal{G}[\mathcal{L}(\mathbf{B})] = \mathcal{L}(\mathcal{G}[\mathbf{B}]))$. We obtain

$$\mathcal{G}_{ij}[\mathcal{L}(B_i(\mathbf{r}))] = \mathcal{G}_{ij}\left(\sum_{\alpha=1}^{4} B_{i,\alpha}\mu_\alpha(\mathbf{r})\right) = \sum_{\alpha=1}^{4} B_{i,\alpha}\mathbf{k}_\alpha. \tag{4.9}$$

Hence, the linear estimators of the divergence and the curl of the field $\mathbf{B}$ are

$$\mathcal{L}(\nabla \cdot \mathbf{B}(\mathbf{r})) = \sum_{\alpha=1}^{4} \mathbf{k}_\alpha \cdot \mathbf{B}_\alpha \quad \text{and} \quad \mathcal{L}(\nabla \times \mathbf{B}(\mathbf{r})) = \sum_{\alpha=1}^{4} \mathbf{k}_\alpha \times \mathbf{B}_\alpha \tag{4.10}$$

The divergence and the curl of the field $\mathbf{B}$ are a combination of the diagonal element and out-of-diagonal element of $\mathcal{G}[\mathcal{L}(\mathbf{B}(\mathbf{r}))]$ respectively.

Magnetic field curvature computation The gradient estimation based on the reciprocal vectors can be used to compute the magnetic tension and the magnetic curvature starting from the gradient matrix of the magnetic field $\nabla\mathbf{B}$. The magnetic tension $\mathbf{T} = \mathbf{B} \cdot \nabla\mathbf{B}$ can be written as

$$\mathbf{T} = \mathbf{B} \cdot \nabla \mathbf{B} = B^2\, \mathbf{b} \cdot \nabla \mathbf{b} + \mathbf{b}(\mathbf{B} \cdot \nabla B) = \mathbf{T}_\perp + \mathbf{T}_{||} \tag{4.11}$$

where $\mathbf{b} = \mathbf{B}/|\mathbf{B}|$. The term $\mathbf{T}_\perp$ is proportional to the magnetic field curvature.

Current Density Estimation: The Curlometer Technique

The curlometer is a multi-spacecraft technique that combines measurements at the four vertexes of the tetrahedron and provides the estimation of the current density $\mathbf{j}$ by means of Ampère law $\nabla \times \mathbf{B} = \mu_0 \mathbf{j}$ (Dunlop et al. 2002; Robert et al. 1998). The displacement current is neglected since non relativistic plasma are considered. In order to apply this method, all the hypothesis listed at the beginning of the section are supposed to hold. Hence, the magnetic field is supposed to vary linearly within the tetrahedron volume and, as a result, the current density has to be constant there. The current density can be computed using the barycentric coordinates method described in Sect. 4.2.2 which allows to extract estimates of $\nabla \times \mathbf{B}$ and $\nabla \cdot \mathbf{B}$. In this section, however, we report the approach that has been presented in Dunlop et al. (2002). The integral definition of $\nabla \times \mathbf{B}$ is given by Stokes theorem

$$\int_\Sigma \mathbf{j} \cdot d\mathbf{s} = \frac{1}{\mu_0} \oint_{\partial\Sigma} \mathbf{B} \cdot d\mathbf{l} \tag{4.12}$$

from which an estimation of the current density averaged over the tetrahedron $\mathbf{j}_{av}$ can be obtained

$$\mathbf{j}_{\beta\gamma\alpha,av} \cdot (\mathbf{r}_{\beta\alpha} \times \mathbf{r}_{\gamma\alpha}) = \frac{1}{\mu_0}(\mathbf{B}_{\beta\alpha} \cdot \mathbf{r}_{\gamma\alpha} - \mathbf{B}_{\gamma\alpha} \cdot \mathbf{r}_{\beta\alpha}) \tag{4.13}$$

where $\mathbf{r}_{\beta\alpha} = \mathbf{r}_\beta - \mathbf{r}_\alpha$, $\mathbf{B}_{\beta\alpha} = \mathbf{B}_\beta - \mathbf{B}_\alpha$ and $\alpha, \beta, \gamma = 1 \ldots 4$ correspond to the four vertexes of the tetrahedron. In Eq. (4.13) the spacecraft at location α is used as reference but we can apply the same formula to three independent faces of the tetrahedron cycling the indexes α, β, γ. In the end, we obtain the expressions for three components of the current density $\mathbf{j}_{\beta\gamma\alpha,av}$ along three directions, each direction is perpendicular to a tetrahedron face. Note that the fourth face, not involving spacecraft α, represents a redundant estimate but it can be used for consistency check. Since the vector defining the face is known by $\mathbf{r}_{\beta\alpha}$ and $\mathbf{r}_{\gamma\alpha}$ the currents normal to three faces can easily be projected into a Cartesian coordinate system (Fig. 4.7).

Uncertainty estimation of the curlometer technique A standard way to establish the measurement performance of the curlometer technique over a set of data is to estimate $\nabla \cdot \mathbf{B}$. Since the magnetic field is a solenoidal field, $\nabla \cdot \mathbf{B} \neq 0$ implies the presence on non-negligible nonlinear contribution to $\mathbf{B}$ that are not taken into account by the curlometer. The divergence of the magnetic field can be computed as follows

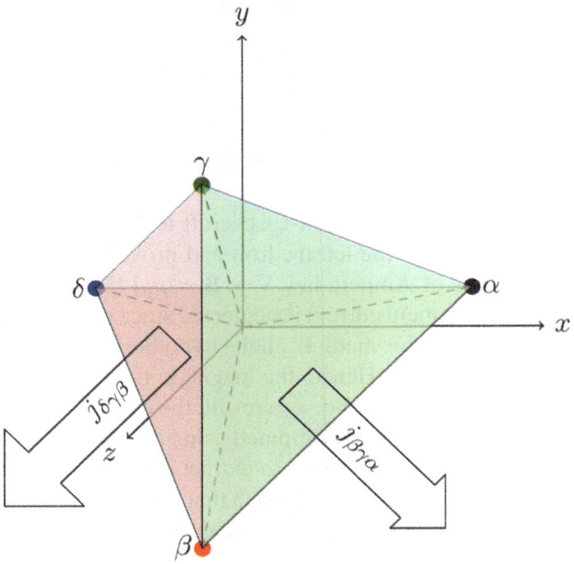

Fig. 4.7 Schematic representation of the current densities computed with the curlometer method

$$\langle\nabla\cdot\mathbf{B}\rangle_{av}|\mathbf{r}_{\beta\alpha}\cdot(\mathbf{r}_{\gamma\alpha}\times\mathbf{r}_{\delta\alpha})| = \left|\sum_{cyclic}\mathbf{B}_{\beta\alpha}\cdot(\mathbf{r}_{\gamma\alpha}\times\mathbf{r}_{\delta\alpha})\right| \tag{4.14}$$

where $\langle\nabla\cdot\mathbf{B}\rangle_{av}$ is the differential estimate of $\nabla\cdot\mathbf{B}$ for the tetrahedron.

Usually, when $\langle\nabla\cdot\mathbf{B}\rangle_{av}/|\mathbf{j}_{av}| \ll 1$ we conclude that the estimation of $\mathbf{j}$ is good. More than providing a quantitative value to the uncertainty of the current density, the field divergence acts as an estimator and it can only indicate whether the estimated current is not reliable. Other sources of error, beyond the non-linearity of the magnetic field, are the current density relative measurement error $\delta j/|\mathbf{j}_{av}|$ due to uncertainties of the magnetic field and spacecraft positions and the *geometrical error* which takes into account the sensitivity to the spacecraft formation. Timing errors are usually relatively unimportant.

Timing Method

A planar discontinuity is assumed to move with a constant velocity V_{CS} along the direction normal to the discontinuity $\mathbf{n}$ (Fig. 4.8). Each spacecraft encounters the layer at a different time t_α. The spacecraft locations are indicated with $\mathbf{r}_\alpha$ where $\alpha = 1\ldots4$ for MMS (or Cluster). For convenience, and without loss of generality, we choose the origin of our coordinate system to be the barycentre of the spacecraft constellation, so that $\sum_\alpha \mathbf{r}_\alpha = 0$.

Since we assumed a uniform motion we can write

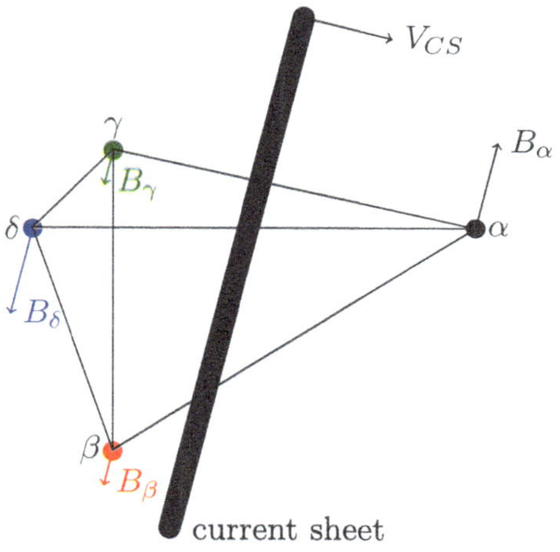

Fig. 4.8 Schematic representation of MMS tetrahedron encountering a planar discontinuity moving with a constant velocity V_{CS}

$$(\mathbf{r}_\alpha - \mathbf{r}_4) \cdot \mathbf{n} = V_{CS}(t_\alpha - t_4) \tag{4.15}$$

where spacecraft 4 has been arbitrary chosen as the reference and $\alpha = 1, 2, 3$. We define $\mathbf{m} = \mathbf{n}/V_{CS}$. Then, the linear system to be solved is

$$\begin{pmatrix} \mathbf{r}_1 - \mathbf{r}_4 \\ \mathbf{r}_2 - \mathbf{r}_4 \\ \mathbf{r}_3 - \mathbf{r}_4 \end{pmatrix} \mathbf{m} = \begin{pmatrix} t_1 - t_4 \\ t_2 - t_4 \\ t_3 - t_4 \end{pmatrix}$$

In order to solve this system the matrix containing the positions has to be non-singular. This condition is satisfied if and only if the four spacecraft are not co-planar. To determine the normal direction we minimize the quantity S:

$$S = \sum_{\alpha=1}^{4} \sum_{\beta=1}^{4} [(\mathbf{r}_\alpha - \mathbf{r}_\beta) \cdot \mathbf{m} - (t_\alpha - t_\beta)]^2. \tag{4.16}$$

Taking into account $\sum_\alpha \mathbf{r}_\alpha = 0$ and defining $t_0 = \frac{1}{4} \sum_\alpha t_\alpha$, S becomes

$$S = \sum_{\alpha=1}^{4} [\mathbf{r}_\alpha \cdot \mathbf{m} - (t_\alpha - t_0)]^2. \tag{4.17}$$

Therefore, we minimize S by putting $\partial S/\partial m_k = 0$ and we obtain

$$\sum_{\alpha=1}^{4} [r_{\alpha,j} m_j - (t_\alpha - t_0)] r_{\alpha,k} = 0. \tag{4.18}$$

Hence,

$$m_j \mathcal{R}_{jk} = \frac{1}{4} \sum_{\alpha=1}^{4} t_\alpha r_{\alpha,k} \quad \text{where} \quad \mathcal{R}_{jk} = \frac{1}{4} \sum_{\alpha=1}^{4} r_{\alpha,j} r_{\alpha,k}. \tag{4.19}$$

The tensor $\mathcal{R}_{jk}$ has to be inverted to determine $\mathbf{m}$. The inverse tensor $(\mathcal{R}_{jk})^{-1}$ can be written as

$$(\mathcal{R}_{jk})^{-1} = 4 \left(\sum_{\alpha=1}^{4} r_{\alpha,j} r_{\alpha,k} \right)^{-1} = 4 \sum_{\alpha=1}^{4} k_{\alpha,j} k_{\alpha,k} \tag{4.20}$$

where $\mathbf{k}_\alpha$ are the reciprocal vectors of the tetrahedron (see Sect. 4.2.2). After calculations, we find that the sought $\mathbf{m}$ vector can be written in terms of the reciprocal vectors

$$\mathbf{m} = \sum_\alpha t_\alpha \mathbf{k}_\alpha \tag{4.21}$$

In this section, I followed the unconstrained homogeneous least squares method (Harvey 1998) to provide an expression for $\mathbf{m}$, as well as for $V_{CS} = 1/|\mathbf{m}|$ and $\mathbf{n} = \mathbf{m}/|\mathbf{m}|$. However, this method is equivalent to the linear barycentric method described in Chanteur (1998), Chanteur and Harvey (1998). The equivalence is demonstrated in Chanteur and Harvey (1998), pp. 372–373.

Crossing times estimation The accuracy of the results produced by timing methods depends critically on the determination of the times t_α corresponding to the center of the crossing of each spacecraft. The approach that has been used throughout this Thesis is to determine the time lag between the spacecraft traversals of the discontinuity by cross correlation of corresponding time series. An alternative approach, used e.g. by Haaland et al. (2004) is to fit a hyperbolic tangent curve to the time series of the measured magnetic field component along the maximum variance direction obtained with MVA by a least-squares procedure and then to use the center time and the duration of the fitted curve.

Minimum Directional Derivative (or Difference, MDD) Method

A multi spacecraft method that allows to compute a coordinate frame local to a discontinuity is the Minimum Directional Derivative (or Difference, MDD) method (Shi et al. 2005). It is also able to provide information about the dimensionality of the structure. The MDD technique is based on the analysis of the matrix $L = (\nabla\mathbf{B})(\nabla\mathbf{B})^T$ which gives the maximum, intermediate and minimum eigenvalues $\lambda_1, \lambda_2, \lambda_3$ and the associated eigenvectors $\mathbf{v}_1, \mathbf{v}_2, \mathbf{v}_3$. It can be demonstrated that the three eigenvalues represent the maximum, intermediate and minimum values of $(\mathbf{n} \cdot \nabla\mathbf{B})^2$ ($\mathbf{n}$ is the direction normal to the boundary). Hence, the three eigenvalues can be used as indicators of the dimensionality of the structure, since they identify directions along which $(\mathbf{n} \cdot \nabla\mathbf{B})^2$ has its minimum, intermediate and maximum value.

The main difference between the MVA and the MDD method is that with the MDD method the local frame is computed at every point of the time series allowing to estimate the temporal variation of the directions. Furthermore, the MDD method can directly determine the invariant axis for a quasi-2D structure and find the normal direction to a quasi-1D structure, as demonstrated using Cluster observations and simulations (Shi et al. 2005).

In particular, if $\lambda_1 \simeq \lambda_2 \simeq \lambda_3$ we can infer that the analyzed structure is a 3D structure. If $\lambda_1, \lambda_2 \gg \lambda_3$ we are observing a quasi-2D structure with the invariant direction along $\mathbf{v}_3$. Finally, if $\lambda_1 \gg \lambda_2, \lambda_3$ the structure is quasi-1D and the invariant plane is defined by the directions ($\mathbf{v}_2$,$\mathbf{v}_3$).

Magnetic Field Topology Analysis: The Poincaré and the FOTE Method

In this section I describe an analytic method that allows to find magnetic nulls using magnetic field measurements from the four MMS spacecraft. A standard method to look for magnetic nulls is the Poincaré Index (PI) method (Greene 1992). The PI method assumes odd value when an odd number of magnetic nulls in enclosed in the considered volume (the tetrahedron volume in our case) and an even value when an even number of null points is enclosed. Since the length of the tetrahedron side is usually small compared to the analyzed structures, it is extremely unlikely than more than one null is enclosed inside the volume defined by the spacecraft constellation. Therefore, PI assumes either the value 1 (a null is found), either the value 0 (no null is present). The PI method has been extensively applied to Cluster data (Eriksson et al. 2015).

A second method to identify a magnetic null is the First Order Taylor Expansion (FOTE) method (Fu et al. 2015, 2016). The FOTE method is able to find a null even if it is located outside the spacecraft tetrahedron. This is particularly suitable for MMS data since the minimum spacecraft separation is 10 times smaller than the minimum separation of Cluster tetrahedron. For this reason, the FOTE method is applied to MMS data in Chap. 7. Indeed, since the inter-spacecraft separation is really small in that case ($\sim$6 km), the magnetic null is located outside the tetrahedron. Hence, the PI method would not find a null.

The FOTE method lays on the hypothesis that the magnetic field can be linearized near the position of the null:

$$\mathbf{B} \approx \mathbf{B}(\mathbf{r}_{null}) + \nabla\mathbf{B} \cdot (\mathbf{r} - \mathbf{r}_{null}). \tag{4.22}$$

Then, since $\mathbf{B}(\mathbf{r}_{null}) \approx 0$, $\mathbf{B} \approx \nabla\mathbf{B} \cdot (\mathbf{r} - \mathbf{r}_{null})$ and $d\mathbf{r} = \mathbf{r} - \mathbf{r}_{null}$ may be derived as follows

$$d\mathbf{r} \approx (\nabla\mathbf{B})^{-1}\mathbf{B}. \tag{4.23}$$

It is worth to note that in this way a null is always found. Depending on the case of study, a threshold on $d\mathbf{r}$ has to be imposed in order to eliminate the non-physical nulls. (Fu et al. 2015) suggest not to trust nulls that are more than 1 d_i away from the spacecraft. (Fu et al. 2015) show that the FOTE method reproduces the results of the PI method. In other words, once a proper threshold is imposed, the results from the two methods agree. Since the case study discussed in Chap. 7 involves electron rather than ion scales, the threshold is strengthened and a null is considered as physically meaningful only if the null-spacecraft distance is $<\frac{1}{4}\ d_i$.

In principle, the FOTE method is also able to identify the null type which is imposed by the characteristics of the matrix $(\nabla \mathbf{B})^{-1}$ (Greene 1988; Parnell et al. 1996). In particular, as demonstrated in Parnell et al. (1996), each field line $\mathbf{r}(k)$ may be written in terms of the eigenvalues λ_1, λ_2, λ_3 and eigenvectors $\mathbf{x}_1$, $\mathbf{x}_2$, $\mathbf{x}_3$ of the matrix $(\nabla \mathbf{B})^{-1}$:

$$\mathbf{r}(k) = A \exp{(\lambda_1 k)}\mathbf{x}_1 + B \exp{(\lambda_2 k)}\mathbf{x}_2 + C \exp{(\lambda_3 k)}\mathbf{x}_3 \tag{4.24}$$

where k is an arbitrary parameter and A, B and C are constant along a field line. Note that the constraint $\nabla \cdot \mathbf{B} = 0$ implies that the trace of $(\nabla \mathbf{B})^{-1}$ has to be zero.

Hence, if all the eigenvalues of $(\nabla \mathbf{B})^{-1}$ are real, there is one of opposite sign to the other two. If we take $\lambda_1 < 0$, $\lambda_2 < 0$ and $\lambda_3 > 0$ it implies that for $k \rightarrow +\infty$

$$\mathbf{r}(k) = C \exp{(\lambda_3 k)}\mathbf{x}_3 \tag{4.25}$$

while for $k \rightarrow -\infty$

$$\mathbf{r}(k) = A \exp{(\lambda_1 k)}\mathbf{x}_1 + B \exp{(\lambda_2 k)}\mathbf{x}_2. \tag{4.26}$$

Hence, the field lines directed towards the null lie in the plane defined by the span of $(\mathbf{x}_1, \mathbf{x}_2)$ while the field lines are directed away from the null along $\mathbf{x}_3$. The $(\mathbf{x}_1, \mathbf{x}_2)$ plane is usually called the *fan* and the $\mathbf{x}_3$ direction is the *spine* of the null (Fig. 4.9(left)). A null as the one shown in Fig. 4.9(left) is called a A-type null (or A radial null), while a null with $\lambda_1 > 0$, $\lambda_2 > 0$ and $\lambda_3 < 0$ is a B-type null.

Another configuration presents two complex and one real eigenvalues. To preserve the solenoidality of the magnetic field, the eigenvalues will be of the type $\lambda_1 = \alpha + i\beta$, $\lambda_2 = \alpha - i\beta$, $\lambda_3 = -2\alpha$. In this case, field lines in the fan plane $(\mathbf{x}_1, \mathbf{x}_2)$ will be spirals while the spine lies in the direction of the eigenvector $\mathbf{x}_3$. If α is positive, the null type is As (A spiral, Fig. 4.9(right)), if $\alpha < 0$ the null type is Bs.

These nulls have a 3D structure that can degenerate in a 2D structure in some conditions. In particular, an A or B type null with an eigenvalue which is zero degenerates to an X type null; an As or Bs type null degenerates to an O type null. Observational data do not provide eigenvalues that are exactly zero. Hence, a null can be considered as 2D when e.g. $\lambda_1 \ll \lambda_2, \lambda_3$ (A and B type) and $\alpha \ll \beta$ (As and Bs type).

As an example, the "empirical" criteria provided in Fu et al. (2015) to establish whether a null can be considered as 2D is reported. In particular, an A or B type null can be simplified to an X type if $\min(|\lambda|) < \frac{1}{4} \max(|\lambda|)$ and an As or Bs type

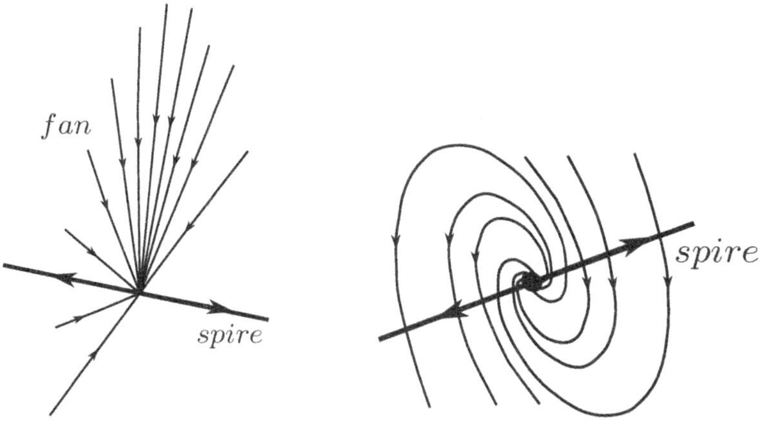

Fig. 4.9 Schematic representation of a A-type radial magnetic null (left) and of a As-type spiral magnetic null (right)

null can be considered as an O type 2D null if $\min(|\Re(\lambda)|) < \frac{1}{4}\max(|\Im(\lambda)|)$. These thresholds need to be adapted to the specific case.

It is not straightforward to associate an uncertainty to the results obtained with the FOTE method. As mentioned above, the first fundamental hypothesis is that the magnetic field changes linearly in the region crossed by the spacecraft. Usually, it is reasonable to suppose that this is the case when the region is relatively small compared to observed structure. For this reason, as mention above, a "distance of validity" (from the spacecraft) is introduced ($\sim 1d_i$ or less). Furthermore, Fu et al. (2015) proposes two quantities to estimate the uncertainty of the method. Since $\nabla \cdot \mathbf{B} = \lambda_1 + \lambda_2 + \lambda_3 = 0$, $\eta = |\nabla \cdot \mathbf{B}|/|\nabla \times \mathbf{B}|$ (already used for the curlometer technique (Dunlop et al. 2002)) and $\xi = |\lambda_1 + \lambda_2 + \lambda_3|/|\lambda|_{max}$ can give an estimation of the error. The two quantities follow the same trend but ξ results to be more sensitive to changes in the $(\nabla \mathbf{B})^{-1}$ matrix.

Furthermore, Eriksson et al. (2015) presented a method to estimate the reliability of the null type identification by comparing the observed local fluctuations of the magnetic field for the particular analysed event with the minimum theoretical disturbances required to alter the null type. A null type can be altered so that it changes type between a A-type null and a B-type or between a spiral and a radial null. Eriksson et al. (2015) performed a statistical study in the Earth's magnetotail and found that relatively small fluctuations in the magnetic field are sufficient to alter the null type. However, for at least 70% of the magnetic nulls in the data set used by Eriksson et al. (2015) the type identification is reliable.

References

D.N. Baker et al., Space Sci. Rev. **199**, 545 (2016)

J.L. Burch, T.E. Moore, R.B. Torbert, B.L. Giles, Space Sci. Rev. **199**, 5 (2016)

C.W. Carlson, J.P. McFadden, P. Turin, D.W. Curtis, A. Magoncelli, Space Sci. Rev. **98**, 33 (2001)

N.A. Case, J.A. Wild, J. Geophys. Res. **117**, A02101 (2012)

G. Chanteur, in ed. by G. Paschmann, P. Daly. *Analysis Methods for Multi-Spacecraft Data* (ISSI Scientific Reports Series, ESA/ISSI, Bern, 1998), vol. 1, p. 349

G. Chanteur, C.C. Harvey, in ed. by G. Paschmann, P. Daly. *Analysis Methods for Multi-Spacecraft Data* (ISSI Scientific Reports Series, ESA/ISSI, Bern, 1998), vol. 1, p. 371

M.W. Dunlop, A. Balogh, K.-H. Glassmeier, P. Robert, J. Geophys. Res. **107**, 23–1 (2002)

R.E. Ergun et al., Space Sci. Rev. **199**, 167 (2016)

E. Eriksson, A. Vaivads, YuV Khotyaintsev, V.M. Khotyayintsev, M. André, Geophys. Res. Lett. **42**, 6883 (2015)

C.P. Escoubet, R. Schmidt, M.L. Goldstein, Space Sci. Rev. **79**, 11 (1997)

H.S. Fu et al., J. Geophys. Res. **120**, 3758 (2015)

H.S. Fu et al., J. Geophys. Res. **121**, 1263 (2016)

S.A. Fuselier, W.S. Lewis, C. Schiff, R.E. Ergun, J.L. Burch, S.M. Petrinec, K.J. Trattner, Space Sci. Rev. **199**, 77 (2016)

J.M. Greene, J. Geophys. Res. **93**, 8583 (1988)

J.M. Greene, J. Comput. Phys. **98**, 194 (1992)

S.E. Haaland et al., Ann. Geophys. **22**, 1347 (2004)

C.C. Harvey, in ed. by G. Paschmann, P. Daly. *Analysis Methods for Multi-Spacecraft Data* (ISSI Scientific Reports Series, ESA/ISSI, Bern, 1998), vol. 1, p. 307

P.-A. Lindqvist et al., Space Sci. Rev. **199**, 137 (2016)

B. Mailyan, C. Munteanu, S. Haaland, Ann. Geophys. **26**, 2383 (2008)

C.E. Parnell, J.M. Smith, T. Neukirch, E.R. Priest, Phys. Plasmas **3**, 759 (1996)

C. Pollock et al., Space Sci. Rev. **199**, 331 (2016)

P. Robert, Patrick, M.W. Dunlop, A. Roux, G. Chanteur, in ed. by G. Paschmann, P. Daly. *Analysis Methods for Multi-Spacecraft Data* (ISSI Scientific Reports Series, ESA/ISSI, Bern, 1998), vol. 1, p. 395

C.T. Russell, B.J. Anderson, W. Baumjohann, K.R. Bromund, D. Dearborn, D. Fischer et al., Space Sci. Rev. **199**, 189 (2016)

Q.Q. Shi et al., Geophys. Res. Lett. **32**, L12105 (2005)

B.U.Ö. Sonnerup, M. Scheible, in ed. by G. Paschmann, P. Daly. *Analysis Methods for Multi-Spacecraft Data* (ISSI Scientific Reports Series, ESA/ISSI, Bern, 1998), vol. 1, p. 185

R.B. Torbert et al., Space Sci. Rev. **199**, 105 (2016)

Chapter 5
Numerical Models for Astrophysical Plasmas

This chapter is meant to briefly present the different numerical models that are commonly used to study astrophysical plasmas. Section 5.1 recalls the kinetic description of a plasma eventually deriving the Vlasov equation which describes the evolution of the distribution function in the collisionless limit. Section 5.2 discusses the different approaches that can be used to model plasma processes. The fluid (Magneto-Hydro-Dynamics, MHD) model is detailed and the role of the Hall term of Ohm's law in promoting fast reconnection is discussed in Sect. 5.2.1. Then, the hybrid model is discussed (Sect. 5.2.2), focusing in particular on the hybrid-Vlasov-Maxwell (HVM) model presented in Valentini et al. (2007). This code is especially significant in the frame of this Thesis since the fully kinetic Vlasov-Darwin code (the ViDA code) that will be extensively described in Chap. 8 has been developed starting from this hybrid code. The last part of the chapter, Sect. 5.2.3, is devoted to kinetic codes. A brief description of Particle-In-Cell (PIC) codes is followed by a subsection about the state of the art of full-kinetic Vlasov codes aiming to provide the context for the ViDA code presented in Chap. 8.

5.1 Kinetic Description

The most complete description of a plasma would include the knowledge of the position and momentum for each particle of the plasma at each time. Since the number of particles of the plasma is too large to allow for a single particle description, a statistical approach is usually used. The statistical description is theoretically achieved through the distribution function F_N

$$F_N(\mathbf{r}_1 \dots \mathbf{r}_N, \mathbf{v}_1 \dots \mathbf{v}_N, t) \tag{5.1}$$

G. Cozzani, *Microphysics of Magnetic Reconnection in Near-Earth Space*, Springer Theses, https://doi.org/10.1007/978-3-030-56142-0_5

where N is the number of particles and F_N is the density of probability that the system is found in a point of the $6N$-dimensional phase space at time t. Usually, $\int F_N\, d\mathbf{r}_1 d\mathbf{v}_1 \ldots d\mathbf{r}_N d\mathbf{v}_N = 1$ is imposed. The evolution of F_N can be obtained with the following hypothesis: the total number of particles in the system is constant, particles of the same species are not distinguishable and the dynamics of the particle is Hamiltonian. The first hypothesis leads to a continuity equation for F_N. Note that in this way the ionization process is not taken into account. Indeed, ionization is usually negligible in the conditions that we are going to analyse.

The third hypothesis is equivalent to say that the Hamiltonian dynamics preserves the volumes in the phase space (Liouville's theorem). With these two premises we obtain that the equation describing the evolution of the distribution function F_N is

$$\frac{dF_N}{dt} = \frac{\partial F_N}{\partial t} + \sum_{i=1\ldots N} \left(\mathbf{v}_i \cdot \frac{\partial F_N}{\partial \mathbf{r}_i} + \mathbf{a}_i \cdot \frac{\partial F_N}{\partial \mathbf{v}_i} \right) = 0 \tag{5.2}$$

where $\mathbf{a}_i$ contains both the acceleration due to interactions with other particles and the one due to external fields.

It is worth to clarify that Eqs. (5.1) and (5.2) are written for a single species in the plasma. If the plasma is composed by electrons and protons there will be two $F_{N,\alpha}$, $\alpha = e, p$ and the total phase space will be given by the cartesian product of the phase space of the single species with a number of particle N_α. The dimension of the total system will be $6N = 6N_e \times 6N_p$. Then, in order to know the evolution of the whole system, Eq. (5.2) needs to be solved for each species. In the following, in order to keep the notation as simple as possible, a single F_N will be considered.

Now, the microscopic density of particles in a point of the $6N$ dimensional phase space (i.e. the number of particle that are located at ($\mathbf{r}$, $\mathbf{r} + d\mathbf{r}$) with velocity ($\mathbf{v}$, $\mathbf{v} + d\mathbf{v}$) divided by the volume $d\mathbf{r}d\mathbf{v}$) is given by

$$n^{micro}(\mathbf{r}, \mathbf{v}) = \sum_{i=1\ldots N} \delta(\mathbf{r} - \mathbf{r}_i(t))\delta(\mathbf{v} - \mathbf{v}_i(t)). \tag{5.3}$$

The macroscopic quantities corresponds to the statistical average of the corresponding microscopic quantities and they are weighted with the density of probability F_N. Hence, the macroscopic density of particles in the phase space is

$$n^{macro}(\mathbf{r}, \mathbf{v}) = n(\mathbf{r}, \mathbf{v}) = \int n^{micro}(\mathbf{r}, \mathbf{v})\; F_N(\mathbf{r}_1 \ldots \mathbf{r}_N, \mathbf{v}_1 \ldots \mathbf{v}_N, t)\; d\mathbf{r}d\mathbf{v} = \tag{5.4}$$

$$\int \sum_{i=1\ldots N} \delta(\mathbf{r} - \mathbf{r}_i(t))\delta(\mathbf{v} - \mathbf{v}_i(t))\; F_N(\mathbf{r}_1 \ldots \mathbf{r}_N, \mathbf{v}_1 \ldots \mathbf{v}_N, t)\; d\mathbf{r}d\mathbf{v}. \tag{5.5}$$

Since the particles are indistinguishable we have

$$n(\mathbf{r}, \mathbf{v}) = N \int \delta(\mathbf{r} - \mathbf{r}_1(t))\delta(\mathbf{v} - \mathbf{v}_1(t))\; d\mathbf{r}_1 d\mathbf{v}_1 \int F_N\; d\mathbf{r}_2 d\mathbf{v}_2 \ldots d\mathbf{r}_N d\mathbf{v}_N \tag{5.6}$$

and then

$$n(\mathbf{r}, \mathbf{v}) = N \int F_N(\mathbf{r} \ldots \mathbf{r}_N, \mathbf{v} \ldots \mathbf{v}_N, t)\, d\mathbf{r}_2 d\mathbf{v}_2 \ldots d\mathbf{r}_N d\mathbf{v}_N. \tag{5.7}$$

The integral in Eq. (5.7) (for fixed t) depends upon $\mathbf{r}$ and $\mathbf{v}$ only and it corresponds to the single particle distribution function $f_1(\mathbf{r}, \mathbf{v})$

$$f_1(\mathbf{r}, \mathbf{v}, t) = V \int F_N(\mathbf{r} \ldots \mathbf{r}_N, \mathbf{v} \ldots \mathbf{v}_N, t)\, d\mathbf{r}_2 d\mathbf{v}_2 \ldots d\mathbf{r}_N d\mathbf{v}_N \tag{5.8}$$

so that $\int f_1(\mathbf{r}, \mathbf{v}, t)\, d\mathbf{r} d\mathbf{v} = V$. The single particle distribution function $f_1(\mathbf{r}, \mathbf{v}, t)$ allow to progress in the kinetic description of the plasma since it is more manageable than F_N which still depends upon the coordinates of N particles in the phase space.

The evolution of f_1 is obtained via the integration of Eq. (5.2). When performing the integration, we need to consider that $F_N(\mathbf{r}_i = \pm\infty) = 0$ (which corresponds to say that there are no particles at $\pm\infty$) and that in a Hamiltonian system the acceleration does not depends upon the corresponding velocity component. This is the case, for example, of the Lorentz force $\mathbf{F} = e\mathbf{v} \times \mathbf{B}$. In the end, we obtain

$$\frac{\partial f_1}{\partial t} + \mathbf{v} \cdot \frac{\partial f_1}{\partial \mathbf{r}} + \mathbf{a}^{ext} \cdot \frac{\partial f_1}{\partial \mathbf{v}} = \frac{N}{m_1 V} \int \frac{\partial \phi(\mathbf{r}, \mathbf{r}_2)}{\partial \mathbf{r}} \frac{\partial f_2}{\partial \mathbf{v}}\, d\mathbf{r}_2 d\mathbf{v}_2 \tag{5.9}$$

where $\mathbf{a}^{ext}$ is the acceleration due to external fields (e.g. gravitational, electromagnetic field ...) while $\partial_{\mathbf{r}}\phi(\mathbf{r}, \mathbf{r}_2)/m_1$ is the acceleration due to the interaction with another particle and that depends on the coordinates of both involved particles.

The two particle distribution function $f_2(\mathbf{r}_a, \mathbf{r}_b, \mathbf{v}_a, \mathbf{v}_b)$ is the joint probability of having a particle in the phase-space point $(\mathbf{r}_a, \mathbf{v}_a)$ and another in $(\mathbf{r}_b, \mathbf{v}_b)$. In a general way, it can be written as

$$f_2(\mathbf{r}_a, \mathbf{r}_b, \mathbf{v}_a, \mathbf{v}_b) = f_1(\mathbf{r}_a, \mathbf{v}_a) f_1(\mathbf{r}_b, \mathbf{v}_b)[1 + g_2(\mathbf{r}_a, \mathbf{r}_b, \mathbf{v}_a, \mathbf{v}_b)] = \tag{5.10}$$

$$= f_1(\mathbf{r}_a, \mathbf{v}_a) f_1(\mathbf{r}_b, \mathbf{v}_b) + G_2(\mathbf{r}_a, \mathbf{r}_b, \mathbf{v}_a, \mathbf{v}_b) \tag{5.11}$$

where g_2 represents the correlation of the two distribution functions. If $g_2 = 0$ (or equivalently $G_2 = 0$) the two particles are statistically independent. Using Eq. (5.11), we can rewrite Eq. (5.9) as

$$\frac{\partial f_1}{\partial t} + \mathbf{v} \cdot \frac{\partial f_1}{\partial \mathbf{r}} + (\mathbf{a}^{ext} + \mathbf{a}^{mean}) \cdot \frac{\partial f_1}{\partial \mathbf{v}} = \frac{N}{m_1 V} \int \frac{\partial \phi(\mathbf{r}, \mathbf{r}_2)}{\partial \mathbf{r}} \frac{\partial G_2}{\partial \mathbf{v}}\, d\mathbf{r}_2 d\mathbf{v}_2 \tag{5.12}$$

where $\mathbf{a}^{mean}$ is the *mean field* acceleration due to the presence of the plasma. Equation (5.12) is the well known Boltzmann equation and the right hand side is the so-called collisional term.

If the collisional term can be neglected, the plasma is collisionless and the equation that describes the evolution of the distribution function is the Vlasov equation

$$\frac{\partial f_l}{\partial t} + \mathbf{v} \cdot \frac{\partial f_l}{\partial \mathbf{r}} + (\mathbf{a}^{ext} + \mathbf{a}^{mean}) \cdot \frac{\partial f_l}{\partial \mathbf{v}} = 0. \tag{5.13}$$

In general, a plasma can be considered as collisionless if the mean free path λ is much larger than the characteristic scale of the system L and, analogously, the collision frequency ν_{coll} is much smaller than the characteristic frequency of the system T^{-1} and the other frequencies involved in the dynamics (e.g. the cyclotron frequency Ω_c or the plasma frequency ω_p of each species). In this context, collisionless means that Coulomb interactions between two particles do not significantly alter the dynamics of the charged particles compared to the effect of the electromagnetic fields that are collectively generated. This is often the case in astrophysical plasmas since they are sufficiently dilute and warm.

The Vlasov equation can be interpreted in a intuitive way. The number of particles within a volume in the phase space $d\mathbf{r}d\mathbf{v}$ changes when particles moving with a certain velocity change their location in space and particles that undergo the action of a force change their position in the velocity space. The force are composed by external forces and the forces given by the mean field, as mentioned above. Note that the number of particles in a volume can be also change via ionization/recombination processes. However, also these processes are negligible in the collisionless and fully ionized plasmas which are discussed in this Thesis. Within this context, the Vlasov equation for each particle species, self-consistently coupled to Maxwell equations, provides a complete description of the system dynamics.

5.2 Numerical Models for Astrophysical Plasmas

It is extremely hard to deal with the Vlasov-Maxwell system of equations analytically, even when we want to solve a simplified problem. For this reason, numerical simulations are presently the most used resource for modeling collisionless plasma.

However, when we want to simulate a collisionless plasma – and we have realistic computational resources at our disposal – we need to perform some approximations. Indeed, it is not possible to follow the dynamics of every particle that composes the system that we want to model. For this reason, a number of approaches with a different number of underlying hypothesis have been developed. This section contains an overview of the most used simulation approaches to model plasma processes in space (fluid, hybrid and kinetic, both PIC and Vlasov). The aim of the section is summarizing the underlying hypothesis and the limitations of the different schemes, together with an overview of the physical processes that can be studied with each scheme. It is worth to underline that the kinetic codes are the only ones providing details of electron dynamics. Hence, these approaches are crucial in the context of this Thesis. In particular, Sect. 5.2.3 focuses on the Vlasov codes since part of this Thesis is devoted to the testing and the use of the newly developed fully-kinetic Eulerian Vlasov-DArwin algorithm (ViDA), as detailed in Chap. 8.

When modeling the near-Earth plasmas, there are some common hypothesis shared by all the approaches. As already mentioned, the plasmas are collisionless and magnetized. Moreover, they can be considered as non-relativistic; the effects of gravity are negligible and quasi-neutrality is fulfilled.

5.2.1 Fluid Codes

In general, the larger is the modeled system, the fewer are the details that can be taken into account with simulations. Hence, when we want to treat large scale phenomena (e.g. involving the whole magnetopause) the most suited model would be a Magneto-Hydro-Dynamics (MHD) or fluid code. Within these models, the plasma dynamics is described with (single or multi) fluid equations which are theoretically obtained by computing the momenta of Eq. (5.13).

If the plasma is composed by electrons and protons (as it is often the case in near-Earth space), the integration of Eq. (5.13) of each species leads to the two fluid model. In this case, the motion of the single particle is completely neglected and the processes that can be reproduced have characteristic spatial and temporal scales which are much larger that the particles scales ($L \gg \rho_p, d_p, T \gg \Omega_{c,p}^{-1}, \omega_{p,p}^{-1}$).

In order to use these models, it is necessary to introduce a fluid closure which generally corresponds to provide a description for the kinetic pressure. Solutions which are often adopted are the adiabatic or the isothermal pressure which are described by polytropic closure of the type $Pn^{-\gamma} = const$ (γ is the polytropic index which is linked to the degrees of freedom ν of the particles $\gamma = \frac{\nu+2}{\nu}$). When $\gamma = 1$ the pressure is isothermal and the pressure tensor is reduced to a scalar $P \propto nT$ where n is the density and T the temperature.

Combining the fluid equations for the two species the MHD model is obtained. In this case, the plasma is seen as a single fluid and a fluid closure for the kinetic pressure is still needed.

The simplest MHD model is the so-called *ideal* MHD for which the Ohm's law is written as follows

$$\mathbf{E} = -\frac{\mathbf{u} \times \mathbf{B}}{c} \tag{5.14}$$

which combined to the Faraday equation leads to

$$\frac{\partial \mathbf{B}}{\partial t} = \nabla \times (\mathbf{u} \times \mathbf{B}) \tag{5.15}$$

where $\mathbf{u} = (m_p\mathbf{u}_p + m_e\mathbf{u}_e)/(m_p + m_e)$ is the bulk velocity of the single fluid plasma. This MHD model is *ideal* in the sense it does not include diffusion nor dissipation. Also, Eq. (5.15) implies that the plasma is *frozen* to the magnetic field at all times (see Chap. 2). An important consequence, in the context of this Thesis, is that the ideal MHD model can not reproduce magnetic reconnection. For this reason, the so-called

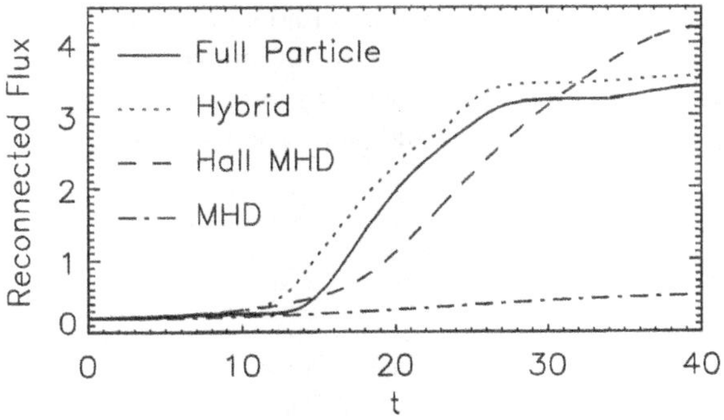

Fig. 5.1 The reconnected magnetic flux versus time from a variety of simulation models as shown in Birn et al. (2001). © 2001 by the American Geophysical Union

resistive MHD and Hall MHD models have been used to perform magnetic reconnection simulations based on the MHD formalism (Birn et al. 2001, and references therein, Lottermoser and Scholar 1997).

The resistive MHD model introduces a resistivity term $\eta \mathbf{j}$ to the right hand side of Ohm's law which allows the magnetic reconnection instability to develop. The resistivity η can be uniform in the simulation box or localized and current-dependent. The Hall MHD model modifies the ideal Ohm's law Eq. (5.14) introducing the Hall term $\propto \mathbf{j} \times \mathbf{B}$. The Hall MHD model is able to describe the demagnetization of ions while the electrons stay frozen to the electric field. Indeed, assuming that at the large MHD scales the dynamics is dominated by protons ($\mathbf{u} \approx \mathbf{u}_p$) we obtain the following Ohm's law

$$\mathbf{E} = -\frac{\mathbf{u}_p \times \mathbf{B}}{c} + \frac{\mathbf{j} \times \mathbf{B}}{nec} = -\frac{\mathbf{u}_e \times \mathbf{B}}{c} \tag{5.16}$$

which indicates that the electrons are magnetized.

It has been shown by Birn et al. (2001) in the context of the Geospace Environmental Modeling (GEM) magnetic reconnection challenge that the inclusion of the Hall term in Ohm's law is necessary to produce reconnection rates comparable to the one obtained by hybrid and kinetic models (see also Shay et al. 2001). Figure 5.1 shows the reconnected flux of a 2D magnetic reconnection simulation with the same parameters performed by different type of codes. Note that the behaviour of the Hall MHD models is similar to the kinetic codes (which include the Hall effect as well) while the resistive MHD model (with uniform resistivity η) does not reproduce fast reconnection.

Since this comparison has been done on 2D simulations its outcome is not general and it still need to be confirmed in a 3D scenario. However, an intuitive explanation of the fact that fast reconnection is possible only with the inclusion of the Hall term can be provided. Indeed, the Hall term begins to play a role at the ion scales d_p while the terms that allows for reconnection become non negligible at smaller scales. In particular, the pressure term plays a role at sub-ion scales, the electron inertia term plays a role at the electron scale d_e and the relative importance of the resistive term compared to the ideal term in Ohm's law is $\propto R_m^{-1}$, the inverse of the magnetic Reynolds number. This scaling determines the presence of different regions at different scales. At ion scales only the ions are demagnetised while the terms that allows reconnection intervene at smaller scales. Hence, in the Hall region the magnetic field is convected by electrons. The electrons, being lighter than the ions, allow the magnetic field to have a faster dynamics so that fast reconnection finally takes place. If the Hall term is not present, both species are frozen to the magnetic field up to the resistive scales and this leads to a slower reconnection process.

5.2.2 Hybrid Codes

In order to include effects at the ion scales, the MHD (or multi-fluid) model can be extended to hybrid models, in which e.g. the ions are treated kinetically and the electrons are modeled as a fluid. The hybrid scheme is introduced to simplify the description of electrons, in order to eliminate their fast and small scale dynamics. Electrons are usually reduced to a isothermal, isotropic fluid and are often treated as massless. In particular, the hybrid scheme allows to include the terms related to the complete pressure tensor of the ions. On the other side a fluid closure is still needed in order to model the electron pressure. There are several possibilities to define the electron pressure but the most used are the isothermal and the adiabatic closures. Some closures incorporate pressure anisotropy, e.g. the Chew-Goldberger-Low (CGL) model (Chew et al. 1956).

There are several examples of hybrid codes that have been used to investigate plasma processes (in particular at ion scales) in the field of astrophysical plasmas. In particular, hybrid-PIC codes has been used to perform global simulations of the magnetosphere (Karimabadi et al. 2014, and reference therein, Lin et al. 2014). Nowadays, the available computational resources even allow to perform global hybrid-Vlasov simulations of the magnetosphere (von Alfthan et al. 2014), giving a more detailed picture compared to the global MHD or fluid codes.

The hybrid-Vlasov-Maxwell (HVM) code (Valentini et al. 2007) is especially interesting in the frame of this Thesis since the full kinetic Vlasov-Darwin code (the ViDA code) that will be extensively described in Chap. 8 has been developed starting from this hybrid code.

The main assumptions of this formalism include that the characteristic length L and T of the analyzed process are $L \gg \lambda_{D,e}$, $T \gg \omega_{p,e}^{-1}$; the plasma is weakly magnetized ($\Omega_{c,e} \ll \omega_{p,e}$) and the displacement current $\propto \partial \mathbf{E}/\partial t$ is neglected. Neutrality is also assumed ($n_e = n_p = n$) so that $\nabla \cdot \mathbf{E} = 0$. With this hypothesis, Valentini et al. (2007) provide the following generalized Ohm's law

$$\mathbf{E} - \frac{m_e c^2}{4\pi n e^2}\nabla^2\mathbf{E} = -\frac{1}{c}(\mathbf{u}_p \times \mathbf{B}) + \frac{1}{nec}(\mathbf{j} \times \mathbf{B}) - \frac{1}{ne}\nabla P_e + \\ + \frac{1}{ne}\frac{m_e}{m_p}\nabla \cdot \overline{\mathbf{P}}_p + \frac{m_e}{ne^2}\left(\mathbf{j}\mathbf{u}_p + \mathbf{u}_p\mathbf{j} - \frac{\mathbf{j}\mathbf{j}}{ne}\right) \tag{5.17}$$

When the system can be described with the MHD model (the typical wavelength L present in the plasma is much larger than d_p and the typical time scale $T \gg \Omega_{c,p}^{-1}$) Eq. (5.17) reduces to Eq. (5.14) as expected. When the typical wavelength and frequency become comparable to the proton scales, also the Hall term and the electron pressure term have to be taken into account, while the terms proportional to m_e in both right and left hand side of Eq. (5.17) are still negligible. If the characteristic length and time are intermediate between the proton and the electron scales ($d_p > L \gg d_e$, $\Omega_{c,p}^{-1} > T \gg \Omega_{c,e}^{-1}, \omega_{p,e}^{-1}$) no term in Eq. (5.17) can be neglected so that Eq. (5.17) becomes a Helmholtz partial differential equation. In these conditions, dynamics is dominated by electrons (Bulanov et al. 1992).

Finally, the HVM scheme is composed by the following equations:

$$\frac{\partial f_p}{\partial t} + \mathbf{v} \cdot \frac{\partial f_p}{\partial \mathbf{r}} + (\mathbf{E} + \mathbf{v} \times \mathbf{B}) \cdot \frac{\partial f_p}{\partial \mathbf{v}} = 0; \tag{5.18}$$

$$P_e = nk_B T_e \quad \text{if the closure is isothermal;} \tag{5.19}$$

$$\frac{\partial \mathbf{B}}{\partial t} = -c\nabla \times \mathbf{E}; \tag{5.20}$$

$$\nabla \times \mathbf{B} = \frac{c}{4\pi}\mathbf{j}; \tag{5.21}$$

$$\mathbf{j} = ne(\mathbf{u}_p - \mathbf{u}_e); \tag{5.22}$$

$$\mathbf{E} - \frac{m_e c^2}{4\pi n e^2}\nabla^2 \mathbf{E} = -\frac{1}{c}(\mathbf{u}_p \times \mathbf{B}) + \frac{1}{nec}(\mathbf{j} \times \mathbf{B}) - \frac{1}{ne}\nabla P_e + \frac{1}{ne}\frac{m_e}{m_p}\nabla \cdot \overline{\mathbf{P}}_p + \frac{m_e}{ne^2}\left(\mathbf{j}\mathbf{u}_p + \mathbf{u}_p\mathbf{j} - \frac{\mathbf{j}\mathbf{j}}{ne}\right); \tag{5.23}$$

where $f_p(\mathbf{r}, \mathbf{v}, t)$ is the proton distribution function. The integrations of f_p provide the proton moments (number density, current density, pressure ...).

To solve the HVM system of equations, the current advance method (CAM) is coupled to the splitting method (see Sects. 3.1 and 3.2 of Valentini et al. 2007). The CAM method provides the second order numerical solution for the advancement of electric and magnetic fields, while the splitting method (Cheng and Knorr 1976) is a second order scheme in time for the advance of the particle distribution function in phase space (see also Mangeney et al. 2002).

As it has already underlined above, Eq. (5.17) is a Helmholtz equation. For periodic boundary conditions in the physical space it can be solved using standard Fast Fourier Transform algorithms. Even though this computation is heavier compared to solving an algebraic expression (e.g. as in the Hall MHD model), the computational cost is negligible with respect to that needed to integrate the Vlasov equation.

5.2.3 *Kinetic Codes*

The complexity of the description of collisionless magnetic reconnection is largely due to the theoretically challenging parameter regime in the reconnecting current sheet which is characterized by kinetic scales $L_{CS} \sim \rho_e \sim d_e \ll \rho_i \sim d_i$. The scales at play are so small compared to the scales of the global system (especially to the mean free path and the collision frequency) that the coulomb scattering and other collision are be totally negligible within the current sheet. These conditions rule out the possibility to use many of the previously reported descriptions. One of the reasons is the fact that previously reported algorithms have the need of a closure and the used ones are usually not representative of the collisionless physics in the current sheet. Since both species are eventually demagnetised in the Electron Diffusion Region, nothing short of full integration of Vlasov equations for electrons and ions coupled to the electric and magnetic field is appropriate for theoretical analysis. At present, there are two approaches that fulfill the need of a complete kinetic description of a plasma, namely the Particle-In-Cell (PIC) approach and the Vlasov approach.

PIC Codes

Nowadays, Particle-In-Cell (PIC) codes (Birdsall and Langdon 1985) represent the most adopted approach to numerical simulations of plasmas in the framework of the kinetic theory. In particular, the PIC approach is Lagrangian. The PIC scheme is based on the fact that the distribution function whose evolution is described by the Vlasov equation Eq. (5.13) is conserved along the particles trajectories in the phase space. Hence, the distribution function is treated as an ensemble of particles and the system is divided among several *macroparticles* that sample the distribution function. Each macroparticle is centered in point $(\mathbf{r}, \mathbf{v})$ of the phase space and it is provided with a *statistical weight*. The macroparticles that are used to discretize the distribution function move along the characteristics of the Vlasov equation. Hence, the PIC scheme integrates the Vlasov equations along the trajectories of the macroparticles in a continuous phase-space. On the other side, the magnetic and electric field are discretized on a spatial grid and have to be interpolated. The "cell" of the wording Particle-In-Cell refers to the region among grid points where the electromagnetic field is not defined.

In general, PIC codes are frequently adopted because they combine the kinetic approach to relatively affordable computational expenses and to implementation simplicity (Pritchett 2000, and references therein, Markidis et al. 2010; Zeiler et al. 2002; Ricci et al. 2002; Horiuchi and Sato 1999; Hesse et al. 2001; Camporeale and Burgess 2017; Grošelj et al. 2017; González et al. 2019). Tremendous advances in the knowledge of the magnetic reconnection have been done thanks to the PIC approach. Nevertheless, because of the sampling of the distribution function (i.e. the limited number of macroparticles), PIC codes are intrinsically affected by statistical noise. The particle noise induces artificial dissipation and it limits description the high energy tails of the distribution function and of fine resonances.

Vlasov Codes, State of the Art

Full Vlasov simulations, in which both electron and ion Vlasov equations are treated, represent another kinetic approach to plasma physics investigation. This kind of scheme, however, has not been extensively exploited in the past, as it the case for PIC simulations, and even today the examples of full Vlasov collisionless magnetic reconnection are relatively few. This situation is mainly due to the higher complexity of Vlasov codes and to the amount of computational resources that are needed for such schemes. Indeed, while PIC codes treat ordinary differential equations describing macroparticle motion, Vlasov codes directly solve the advection-type partial differential Vlasov equation for each species. Moreover, the distribution function is a 6-dimensional object which depends on both regular space and velocity space which needs to be computed and then stored. For these reasons, the required numerical effort had usually prevented Vlasov codes to be used as much as PIC codes.

Nevertheless, in the last decades there have been examples of development and applications of Vlasov codes (Wiegelmann and Büchner 2001; Schmitz and Grauer 2006a; Umeda et al. 2009, 2012; Delzanno 2015; Roytershteyn and Delzanno 2018). Vlasov codes have been usually applied to problems with lower dimensionality (usually 2D system). Also, the computational capabilities of super-computers have increased lately so that Vlasov codes became relatively spread.

Vlasov codes have been used to investigate magnetic elicity (Wiegelmann and Büchner 2001, 2002), anomalous resistivity (Büchner and Elkina 2005), dynamics of thin magnetopause-like current sheets (Silin and Büchner 2006), the role of the electron pressure tensor during magnetic reconnection (Schmitz and Grauer 2006b), the Kelvin-Helmholtz instability (Umeda et al. 2010a), the structure of diffusion region in magnetic reconnection (Umeda et al. 2010b; Zenitani and Umeda 2014), the development of secondary instabilities in the collisionless Rayleigh-Taylor instability (Umeda et al. 2016).

Indeed, Vlasov schemes have the main advantage not to be affected by statistical noise since the entire distribution function is retained. In particular, they allow to treat resonance problems and phenomena which are related to the high energy tails of the distribution function. Concerning collisionless magnetic reconnection specifically, Vlasov scheme allows for the punctual investigation of the distribution functions (PIC codes usually provide distribution functions averaged over a volume of the simulation box) and precise evaluation of the reconnection electric field with the generalized Ohm's law including anomalous resistivity terms related to wave-particle interaction.

There are several schemes that have been implemented in order to solve the Vlasov equation. Semi-Lagrange schemes were successfully developed for low-dimensional problems (Cheng and Knorr 1976). Another option was also given by transform methods. Then, the *time splitting* method was proposed by Cheng and Knorr (1976). The time splitting method is widely described in Sect. 8.3.1 since it has been adopted in the implementation of the ViDA code presented in Chap. 8. The description of the other methods lies beyond the scope of this Thesis and further information can be found in (Büchner 2007) and references therein.

Together with the integration of the Vlasov equation, the Maxwell equations have to be computed. Vlasov codes also differ depending on the treatment of Maxwell equations which is chosen taking into account the constraint on time step Δt imposed by the Courant-Friedrichs-Lewy (CFL) condition, $\Delta t \lesssim \Delta x/c$ (Courant et al. 1928; Peyret and Taylor 1986). Such condition is due to the fact that Maxwell equations also describe the propagation of waves at the light speed c.

In specific situations, i.e. when the self generated magnetic field can be neglected, this constraint can be avoided using an electrostatic model. In general, the problem has been overcome by neglecting the displacement current in Ampère equation (Wiegelmann and Büchner 2001) or by adopting the Darwin approximation which rules out the transverse light waves (i.e. the fastest waves in the system that propagate at phase speed c) thus significantly relaxing the CFL condition (Schmitz and Grauer 2006a; Mangeney et al. 2002; Pezzi et al. 2019). Since the Darwin approximation has been used for the code ViDA, it will be extensively treated in Sect. 8.1. An alternative method to avoid the CFL condition for light is the use of an implicit Finite Difference Time Domain (FDTD) method for solving Maxwell's equations (Umeda et al. 2009).

In conclusion, Vlasov codes represent a fundamental tool to investigate plasma processes which should not be seen as a substitute of the PIC simulation but rather as a complementary approach. In addition, the development of Vlasov schemes is also crucial because they represent an alternative method that allows to validate the PIC simulations results.

It has also been argued (Büchner 2007) that PIC codes are not less expensive numerically than Vlasov codes when the same level of noise and phase space resolution is required for the two approaches. However, while PIC simulations have proven to provide results in a broad range of problems even if affected by the statistical unavoidable noise, Vlasov codes are inevitably limited by their computational cost.

References

C.K. Birdsall A.B. Langdon, *Plasma Physics Via Computer* (McGraw-Hill, Inc., New York, 1985)

J. Birn et al., J. Geophys. Res. **106**, 3715 (2001)

J. Büchner, in *Advanced Methods for Space Simulations*, ed. by H. Usui, Y. Omura (TERRAPUB, Tokyo, 2007), p. 23

J. Büchner, N. Elkina, Space Sci. Rev. **121**, 237 (2005)

S.V. Bulanov, F. Pegoraro, A.S. Sakharov, Phys. Fluids B **4**, 2499 (1992)

E. Camporeale, D. Burgess, J. Plasma Phys. **83**, 535830201 (2017)

C.Z. Cheng, G. Knorr, J. Comput. Phys. **22**, 330, (1976)

G.F. Chew, M.L. Goldberger, F.E. Low, in *The Boltzmann equation and the one-fluid hydromagnetic equations in the absence of particle collisions*, ed. by S. Chandrasekhar. Proceedings of the Royal

Society A. Mathematical, Physical and Engineering Sciences (The Royal Society Publishing, London, 1956), vol. 236, p. 112
R. Courant, K. Friedrichs, H. Lewy, Math. Ann. **100**, 32 (1928)
G.L. Delzanno, J. Comput. Phys. **301**, 338 (2015)
C.A. González, T.N. Parashar, D. Gomez, W.H. Matthaeus, P. Dmitruk, Phys. Plasmas **26**, 012306 (2019)
D. Grošelj et al., Astrophys. J. **847**, 28 (2017)
M. Hesse, J. Birn, M. Kuznetsova, J. Geophys. Res. **106**, 3721 (2001)
R. Horiuchi, T. Sato, Phys. Plasmas **6**, 4565 (1999)
H. Karimabadi et al., Phys. Plasmas **21**, 062308 (2014)
Y. Lin, X.Y. Wang, S. Lu, J.D. Perez, Q. Lu, J. Geophys. Res. **119**, 7413 (2014)
R.F. Lottermoser, M. Scholar, J. Geophys. Res. **102**, 4875 (1997)
A. Mangeney, F. Califano, C. Cavazzoni, P. Trávníček, J. Comput. Phys. **179**, 495 (2002)
S. Markidis, G. Lapenta, R. Uddin, Math. Comput. Simulat. **80**, 1509 (2010)
R. Peyret, T.D. Taylor, *Computational Methods for Fluid Flow* (Springer, Berlin Heidelberg, 1986)
O. Pezzi et al., J. Plasma Phys. **85**(5), 905850506 (2019)
P.L. Pritchett, IEEE T. Plasma Sci. **28**, 1976 (2000)
P. Ricci, G. Lapenta, J.U. Brackbill, J. Comput. Phys. **183**, 117 (2002)
V. Roytershteyn, G.L. Delzanno, Front. Astron. Space Sci. **5**, 27 (2018)
H. Schmitz, R. Grauer, Phys. Plasmas **13**, 092309 (2006)
H. Schmitz, R. Grauer, J. Comput. Phys. **214**, 738 (2006)
M.A. Shay, J.F. Drake, B.N. Rogers, R.E. Denton, J. Geophys. Res. **106**, 3759 (2001)
I. Silin, J. Büchner, Adv. Space Res. **37**, 1354 (2006)
T. Umeda, K. Togano, T. Ogino, Comput. Phys. Commun. **180**, 365 (2009)
T. Umeda, J. Miwa, Y. Matsumoto, T.K.M. Nakamura, K. Togano, K. Fukazawa, I. Shinohara, Phys. Plasmas **17**, 052311 (2010)
T. Umeda, K. Togano, T. Ogino, Phys. Plasmas **17**, 052103 (2010)
T. Umeda, K. Fukazawa, Y. Nariyuki, T. Ogino, IEEE T. Plasma Sci. **40**, 1421 (2012)
T. Umeda, Y. Wada, Phys. Plasmas **23**, 112117 (2016)
F. Valentini, P. Trávníček, F. Califano, P. Hellinger, A. Mangeney, J. Comput. Phys. **225**, 753 (2007)
S. von Alfthan, D. Pokhotelov, Y. Kempf, S. Hoilijoki, I. Honkonen, A. Sandroos, M. Palmroth, J. Atmos. Sol.-Terr. Phys. **120**, 24 (2014)
T. Wiegelmann, J. Büchner, Nonlin. Processes Geophys. **8**, 127 (2001)
T. Wiegelmann, J. Büchner, Nonlin. Processes Geophys. **9**, 139 (2002)
A. Zeiler, D. Biskamp, J.F. Drake, B.N. Rogers, M.A. Shay, M. Scholer, J. Geophys. Res. **107**, 1230 (2002)
S. Zenitani, T. Umeda, Phys. Plasmas **21**, 034503 (2014)

Chapter 6
The Electron Diffusion Region

The previous chapters are mainly focused on the MHD and two fluid description of magnetic reconnection. In these theoretical frameworks, the EDR is depicted as a "black box" where magnetic field lines change their connectivity and energy is converted from the fields to the plasma. The aim of this Chapter is to discuss how the EDR is operationally defined and identified and then to discuss the processes at electron scale that take place in the EDR including, in particular, the results from the latest MMS observations in synergy with numerical PIC simulations.

6.1 Identification and Characterization of the Electron Diffusion Region

When investigating the EDR, it becomes clear that the definitions that have been proposed within the fluid theoretical frameworks are usually incomplete and operationally imprecise. Indeed, the EDR has been conventionally pictured as a region embedded in the larger ion-scale IDR where electron are demagnetized. Using this definition, EDRs were identified using Polar satellite data at the subsolar magnetopause (Mozer et al. 2003). However, as underlined by Pritchett and Mozer (2009), this EDR definition may be misleading. It has been seen in PIC simulations that locations where electrons are not frozen-in (i.e. where $\mathbf{E} + \mathbf{u}_e \times \mathbf{B} \neq 0$) cover areas much larger than expected and these areas are not embedded in the larger IDR. Moreover, even though regions of electron demagnetization are expected to be very rare in astrophysical plasmas, not all demagnetized layers are electron diffusion regions. A weaker demagnetization (compared to the EDR) can also take place in other thin current layers, for instance at the separatrix.

G. Cozzani, *Microphysics of Magnetic Reconnection in Near-Earth Space*, Springer Theses, https://doi.org/10.1007/978-3-030-56142-0_6

For these reasons, other EDR signatures has been proposed in order to guarantee a correct EDR identification. Notably, a nonzero parallel (to the magnetic field) electric field, electron acceleration, electromagnetic energy conversion, a super-Alfvénic electron outflow, enhanced electron agyrotropy, electron energization and a current sheet thickness of the order of d_e (Pritchett and Mozer 2009; Scudder et al. 2008) were taken into account. The observation of most of these signatures was extremely limited and challenging due to the severe constraint imposed by the temporal resolution of particle instruments before the launch of the MMS mission. In addition, electron demagnetization continued to represent a local, rare and necessary signature to unambiguously identify the EDR. Basically, the idea is not to rely on one single signature but to collect several evidences providing a secure way to identify the EDR. An additional challenge is the need to have operational signatures that could be applied to spacecraft in-situ data.

After the observations of electron demagnetisation reported by Mozer et al. (2003), in situ evidence of an EDR was provided by Scudder et al. (2012) by means of Polar data which showed in particular electron demagnetization and an unusually strong agyrotropy (see Appendix B for the definition of agyrotropy). Then, THEMIS observations at the subsolar magnetopause and in the magnetotail identified the EDR as the site of intense wave activity (Tang et al. 2013) and of electron energization (Oka et al. 2016).

Several PIC simulations have enlightened the electron dynamics in the EDR (Hesse et al. 2014; Bessho et al. 2016; Hesse et al. 2016; Shay et al. 2016) predicting in particular the presence of electrons performing meandering orbits which result in crescent-shaped distribution functions (see also Sect. 2.5). However, distribution functions can have even more complicated shapes (Ng et al. 2012; Shuster et al. 2015) which depends also on the distance from the X-line. Using PIC simulations, Shuster et al. (2015) showed that the electron distribution functions have a triangular shape in vicinity of the X-line while further from the X-line, in the outflow region, the distribution functions exhibit an arc shape and become gradually gyrotropic. The complex shape of the distributions is linked to the electron temperature which is increased by both the action of the reconnection electric field and of the magnetic field normal to the reconnection layer.

The launch of the MMS mission (Burch et al. 2016b) gave new impulse to the investigation of the EDR together with the possibility to confirm the simulation results. Indeed, with the unprecedented resolution of the particle data, MMS allowed to lift some of the constraint that had limited EDR observation in the past, in particular concerning the physical quantities related to electrons (see Chap. 4).

The electron-scale physics in an EDR encounter near a reconnection X-line at the magnetopause has been investigated for the first time with MMS data (Burch et al. 2016a). Since then, 33 EDR encounters have been reported at the magnetopause (Fuselier et al. 2017; Webster et al. 2018; Genestreti et al. 2018). Fuselier et al. (2017) collected the first 12 EDR encounters showing that they took place in a wide variety of conditions in terms of local time, magnetic shear angle between the magnetosheath and magnetospheric magnetic fields and radial distance from the Earth. It is worth to note that the 12 EDR events are selected from a set of 4500 magnetopause crossing, both partial (from the magnetosphere or magnetosheath into the boundary layer) and full (from the magnetosphere into the magnetosheath or vice versa).

The larger set of 33 EDR encounters has been gathered by Webster et al. (2018). Even considering this larger set, the EDRs were encountered in a wide range of conditions. This set of EDR observations, together with simulations of asymmetric reconnection (Hesse et al. 2016; Shay et al. 2016), have provided a list of signatures that are needed to identify the EDR. In particular, the EDR is the site of

1. strong, electron-carried current densities of the order of 1000 nA/m^2;
2. electron agyrotropy $\sqrt{Q}$ with values up to 0.1;
3. parallel electron heating with $T_{e,||}/T_{e,\perp}$ up to $\sim$4;
4. minima of $|B| \sim 5$ nT;
5. large energy conversion from the electromagnetic field to plasmas $\mathbf{E}' \cdot \mathbf{j} \sim 10$ nW/m^3 where $\mathbf{E}' = \mathbf{E} + \mathbf{u}_e \times \mathbf{B}$ (Zenitani et al. 2011);
6. non-negligible parallel (to the magnetic field) electric field;
7. meandering trajectories of electrons resulting in crescent-shaped distribution functions.

Crescent-shaped electron distribution functions are observed in most cases, and they are found on the magnetospheric side of the magnetopause boundary (Burch et al. 2016a), in the electron outflow (Norgren et al. 2016), and in the magnetosheath inflow (Chen et al. 2017).

These signatures, including the crescent-shaped distribution functions, have been observed also during EDR encounters in the Earth's magnetotail (Torbert et al. 2018; Zhou et al. 2019). The EDR has been also indicated as the site of production of whistler waves (Cao et al. 2017).

Note that the parameter involving the minimum of the magnetic field strength is biasing the EDR statistical ensemble towards small guide field configurations. However, the majority of the EDRs encounters that have been reported have relatively small guide field. This might be due to the fact that some of EDR signatures (for instance, the crescent-shaped distributions) are more easily observed when the guide field is small. The only EDR with a strong guide field was reported in the frame of Kelvin-Helmholtz related reconnection at duskside of the magnetopause (Eriksson et al. 2016).

Another signature of the EDR that can be fruitfully used for detecting the EDR in spacecraft data has been identified using PIC simulations by Egedal et al. (2018). Indeed, Egedal et al. (2018) analyse the electron distribution functions of the electrons streaming towards the X-point in the electron diffusion region and they find that they are characterized by strong oblique electron beams. Then, they demonstrate that *"because of the inertia of the beams, they do not follow the direction of the magnetic field as it rotates sharply within the electron-diffusion region. The diffusion region of antiparallel asymmetric reconnection is therefore characterized by electron beams at oblique directions to the local magnetic field."* This behaviour can be seen in the electron distributions observed by MMS.

6.1.1 Evaluation of the Generalized Ohm's Law

A significant effort has been devoted to investigate the generalized Ohm's law Eq. (2.18) to understand which term is responsible for the reconnection electric field at electron scale. This topic has been extensively studied by means of numerical simulations (Hesse et al. 1999, 2016; Divin et al. 2016, and references therein) establishing the role of nongyrotropy of the electron distribution function and of the electron pressure term. However, the comparison between observations and simulations were lacking because of technical limitations in spacecraft data so that only few Ohm's law evaluations were carried out with Cluster data (Vaivads et al. 2004). MMS data provides the full pressure tensor, high resolution particle data and, being a constellation of four spacecraft, the possibility to compute the gradients of physical quantities (see Sect. 4.2.2) which in principle allow to compute all the terms of the generalized Ohm's law. Observations (Torbert et al. 2016) have shown that the contribution of the pressure term is larger that the contribution provided by the electron inertia and that the "residue" is comparable to the electric field itself. However, these results have to be taken with care, since the measured electric field may be affected by large errors. This topic will be discussed in Chap. 7. The analysis of another EDR encounter (Genestreti et al. 2018) confirms that the contribution of the pressure term is larger that the contribution of the electron inertial term. In particular, it appears that both the gyrotropic and agyrotropic pressure contribute to reconnection electric field.

However, the role and the relative importance of the terms of Ohm's law is still largely debated. For instance, for asymmetric reconnection, Hesse et al. (2016) finds that at the X-point the inertia term is the largest contributor to the electric field and Egedal et al. (2018) show that the magnitude of the pressure term and of the electron inertia term are similar but the electron inertia is larger in a region of width $\sim 1\ d_e$ around the X-line.

In addition, the currently available simulations are PIC simulation that always present a level of statistical noise. In Chap. 9 I will make use of a noise free full-Vlasov simulation in order to detail the contribution of the different terms to the Ohm's law in the context of symmetric magnetic reconnection.

6.1.2 Inner and Outer EDR

MMS observations and kinetic simulations focused on the EDR are devoted to resolve the structure of the EDR. For instance, it has become clear that the EDR is actually composed by an inner and an outer EDR. This distinction has been introduced by Karimabadi et al. (2007), Shay et al. (2007) and it accounts for the fact that electrons can be demagnetized at large distances from the X-point, as it has been confirmed by observations (Phan et al. 2007). The inner EDR contains the X-point and a strong current density along the out-of-reconnection-plane direction mainly carried by electrons. Moreover, it is the region where energy is converted from the fields to the plasma so that $\mathbf{E}' \cdot \mathbf{j} > 0$. On the other side, the outer region is characterized by electron jets in which electrons are demagnetized. There, the energy is converted back from the electron jets to the magnetic field ($\mathbf{E}' \cdot \mathbf{j} < 0$) since the electron jets are gradually decelerated and thermalized. Observations of the outer EDR confirmed this picture (Hwang et al. 2017). In addition, crescent-shaped electron distribution functions are observed also in the outer EDR (Hwang et al. 2017).

6.1.3 Energy Conversion

The energy conversion from the magnetic field to the particles of the plasma is one of the main characteristics of the magnetic reconnection process. Investigating where and how energy in converted is crucial to understand collisionless reconnection.

This energy exchange can be expressed with the quantity $\mathbf{E} \cdot \mathbf{j}$. In order to better quantify the energy conversion in the EDR and to identify regions which are physically relevant for the electron energization, Zenitani et al. (2011) proposed an electron-frame energy conversion measure, $\mathbf{E}' \cdot \mathbf{j}$, where $\mathbf{E}' = \mathbf{E} + \mathbf{u}_e \times \mathbf{B}$.

The classical picture of 2D reconnection presents a region of enhanced $\mathbf{E}' \cdot \mathbf{j}$ in the middle of the EDR and in correspondence of the X-point. Simulations and observations have shown that this is not the case when reconnection is asymmetric and when the guide field is present. For enhanced asymmetry, the peak of the energy conversion is located on the magnetospheric side of the boundary (Zenitani et al. 2011; Genestreti et al. 2017) in between the X-point and the stagnation point. On the other hand, if the guide field is higher the energy conversion occurs closer to the electron stagnation point, as the X-point and the stagnation point become closer to each other (Genestreti et al. 2017; Cassak et al. 2017).

Another crucial point to be explored is what is sustaining the energy conversion. (Genestreti et al. 2018) shows that both gyrotropic and agyrotropic terms of the pressure force are involved and that both pressure gradient in the reconnection plane ant out of the reconnection plane are important.

These results compose a picture of the EDR which is extremely complex and rich of processes. The large variety of conditions in terms of guide field asymmetry provide a large set of examples that are still not unified (Webster et al. 2018, see e.g. Fig. 5).

One of the questions that are still open is whether the EDR has a preferred homogeneous or inhomogeneous structure at electron scales and below. EDR has been identified as the site of strong vorticity and of current filamentation (Matthaeus 1982; Che et al. 2011) but observational evidence concerning the structure of the

EDR are still lacking. The four MMS would allow to map the EDR and to enlighten its structure. However, to perform this kind of investigation it is necessary that the four spacecraft are sampling the EDR at the same time and this requirement is not easy to be met, especially at the magnetopause where the electron inertial length $d_e \sim 2$ km. In this Thesis, I present a research work that tries to reply to this long standing question (see Chap. 7).

References

N. Bessho, L.-J. Chen, M. Hesse, Geophys. Res. Lett. **43**, 1828 (2016)
J.L. Burch et al., Science **352**, aaf2939 (2016a)
J.L. Burch, T.E. Moore, R. B. Torbert, B.L. Giles, Space Sci. Rev. **199**, 5 (2016b)
D. Cao et al., Geophys. Res. Lett. **44**, 3954 (2017)
P.A. Cassak et al., J. Geophys. Res. **122**, 11523 (2017)
H. Che, J.F. Drake, M. Swisdak, Nature (London) **474**, 184 (2011)
L.-J. Chen et al., J. Geophys. Res. **122**, 5235 (2017)
A. Divin, V. Semenov, D. Korovinskiy, S. Markidis, J. Deca, V. Olshevsky, G. Lapenta, Geophys. Res. Lett. **43**, 10565 (2016)
J. Egedal et al., Phys. Rev. Lett. **120**, 055101 (2018)
S. Eriksson et al., Geophys. Res. Lett. **43**, 5606 (2016)
S.A. Fuselier et al., J. Geophys. Res. **122**, 5466 (2017)
K.J. Genestreti et al., J. Geophys. Res. **122**, 11342 (2017)
K.J. Genestreti et al., J. Geophys. Res. **123**, 1806 (2018)
M. Hesse, N. Aunai, D. Sibeck, and J. Birn, Geophys. Res. Lett. **41**, 8673 (2014)
M. Hesse, Y.-H. Liu, L.-J. Chen, N. Bessho, M. Kuznetsova, J. Birn, J.L. Burch, Geophys. Res. Lett. **43**, 2359 (2016)
M. Hesse, K. Schindler, J. Birn, M. Kuznetsova, Phys. Plasmas **6**, 1781 (1999)
K.-J. Hwang et al., Geophys. Res. Lett. **44**, 2049 (2017)
H. Karimabadi, W. Daughton, J. Scudder, Geophys. Res. Lett. **34**, L13104 (2007)
W.H. Matthaeus, Geophys. Res. Lett. **9**, 660 (1982)
F.S. Mozer, S.D. Bale, T.D. Phan, J.A. Osborne, Phys. Rev. Lett. **91**, 245002 (2003)
J. Ng, J. Egedal, A. Le, W. Daughton, Phys. Plasmas **19**, 112108 (2012)
C. Norgren et al., Geophys. Res. Lett. **43**, 6724 (2016)
M. Oka, T.D. Phan, M. Øieroset, V. Angelopoulos, J. Geophys. Res. **121**, 1955 (2016)
T.D. Phan, J.F. Drake, M.A. Shay, F.S. Mozer, J.P. Eastwood, Phys. Rev. Lett. **99**, 255002 (2007)
P.L. Pritchett, F.S. Mozer, J. Geophys. Res. **114**, A11210 (2009)
J.D. Scudder, R.D. Holdaway, W.S. Daughton, H. Karimabadi, V. Roytershteyn, C.T. Russell, J.Y. Lopez, Phys. Rev. Lett. **108**, 225005 (2012)
J.D. Scudder, R.D. Holdaway, R. Glassberg, S.L. Rodriguez, J. Geophys. Res. **113**, A10208 (2008)
M.A. Shay et al., Geophys. Res. Lett. **43**, 4145 (2016)
M.A. Shay, J.F. Drake, M. Swisdak, Phys. Rev. Lett. **99**, 155002 (2007)
J.R. Shuster, L.-J. Chen, M. Hesse, M.R. Argall, W. Daughton, R.B. Torbert, N. Bessho, Geophys. Res. Lett. **42**, 2586 (2015)
X. Tang, C. Cattell, J. Dombeck, L. Dai, L.B. Wilson III, A. Breneman, A. Hupach, Geophys. Res. Lett. **40**, 2884 (2013)
R.B. Torbert et al., Geophys. Res. Lett. **43**, 5918 (2016)

R.B. Torbert et al., Science **362**, 1391 (2018)
A. Vaivads et al., Phys. Rev. Lett. **93**, 105001 (2004)
J.M. Webster et al., J. Geophys. Res. **123**, 4858 (2018)
S. Zenitani, M. Hesse, A. Klimas, M. Kuznetsova, Phys. Rev. Lett. **106**, 195003 (2011)
M. Zhou et al., Astrophys. J. **870**, 34 (2019)

Chapter 7
In Situ Spacecraft Observations of a Structured Electron Diffusion Region During Magnetopause Reconnection

This chapter presents observations providing evidence of a structured EDR at the magnetopause. In particular, I present and discuss the results presented in Cozzani et al. (2019) and I include analysis that could not be fully detailed in the published manuscript, namely (i) the detailed determination of the local current sheet coordinate system, (ii) the application of the FOTE method (see Sect. 4.2.2) on the selected data.

As discussed in Chap. 6, significant progress has been done in the characterization of the EDR in terms of electron dynamics, energy conversion from the field to the plasma and characterization of the origin of the reconnection electric field through the evaluation of Ohm's law, by means of MMS data and PIC simulations.

However, the structure of the EDR is still poorly understood and this is especially due to the strict limitations that affected the particle observations in the past. In particular, it is not fully understood whether the EDR has a preferred homogeneous or inhomogeneous structure at electron scales and below. This open question arises for different reasons. On one side, simulations have shown that the EDR is the site of strong vorticity (Matthaeus 1982) and of current filamentation at electron scale that can provide a source of anomalous resistivity, eventually leading to the violation of the frozen-in condition (Che et al. 2011). Then, both spacecraft observations and numerical simulations have shown that the presence of current filaments and of particle-scale structures is occurring in the outflow (Lapenta et al. 2015; Osman et al. 2015; Phan et al. 2016) of magnetic reconnection. Evidence of turbulence have been found in the IDR (Eastwood et al. 2009; Ergun et al. 2017; Graham et al. 2017) where the energy dissipation is not evenly distributed but rather spiky and concentrated in small regions (Fu et al. 2017).

Considering all this, we could ask ourselves whether such complexity could be found also in the EDR.

G. Cozzani, *Microphysics of Magnetic Reconnection in Near-Earth Space*, Springer Theses, https://doi.org/10.1007/978-3-030-56142-0_7

However, the first MMS observations of an EDR (Burch et al. 2016) have been compared to two-dimensional PIC simulations (Shay et al. 2016) and interpreted in terms of a laminar region. Yet, these simulations are two-dimensional, have limited spatial resolution and substantial averaging is commonly performed to reduce noise.

It is only from the last two years that this laminar picture of the EDR has been questioned again. Indeed, three-dimensional PIC simulations (Daughton et al. 2011; Price et al. 2017, 2016), two-dimensional PIC simulations with high spatial resolution (Jara-Almonte et al. 2014), or with low computational noise and larger number of macroparticles (Swisdak et al. 2018) indicate that the EDR can be rather inhomogeneous in electric fields, electron flows, current densities, and energy conversion, with the formation of structures at electron-scale. In addition, recent observations have shown that the presence of standing waves in the EDR leads to oscillatory energy conversion in the EDR (Burch et al. 2018).

However, detailed observations supporting the structuring of the EDR are still lacking. This chapter, covering (Cozzani et al. 2019), shows MMS observations of an EDR encounter at the subsolar magnetopause providing evidence EDR structuring at electron scales.

7.1 MMS Event Selection

In order to study the structure of the EDR, I looked for an event[1] in which all the spacecraft encountered the EDR. The search for events has been carried on by selecting the period in which MMS was in proximity of the subsolar magnetopause (where it is most likely for reconnection to occur) and where the inter-separation among the spacecraft was the smallest. We also made use of the SITL reports (see Sect. 4.1.5) in order to focus on the time intervals that could potentially contain an EDR encounter.

Finally, I selected a magnetopause crossing taking place between 12:05:41.9 and 12:05:44.0 Coordinated Universal Time (UTC)[2] on the 27th of January 2017. During that day, MMS registered many full and partial magnetopause crossings with jets and jet reversals, indicating the possibility of an EDR encounter. MMS stayed mostly in the magnetosheath and in the solar wind and several bow shock crossings were also observed.

At the time of the selected magnetopause crossing, the MMS constellation was located in the subsolar magnetopause region, at (9.3, −1.2, 2.1) R_E in Geocentric Solar Ecliptic (GSE) coordinates (see Appendix A). The MMS location relatively to the magnetopause at around 12:05:00 is shown in Fig. 7.1(right). The mean spacecraft separation was ~6 km, which is the smallest possible for MMS and it corresponds to a few local electron inertial lengths at the magnetopause ($d_e \sim 2$ km). The tetrahedron configuration is shown in Fig. 7.1(left). The quality factor $Q = 0.843$ (see Sect.

[1]The term *event* is commonly used in this context to indicate the chosen set of data.

[2]All the times are in UTC when not differently specified.

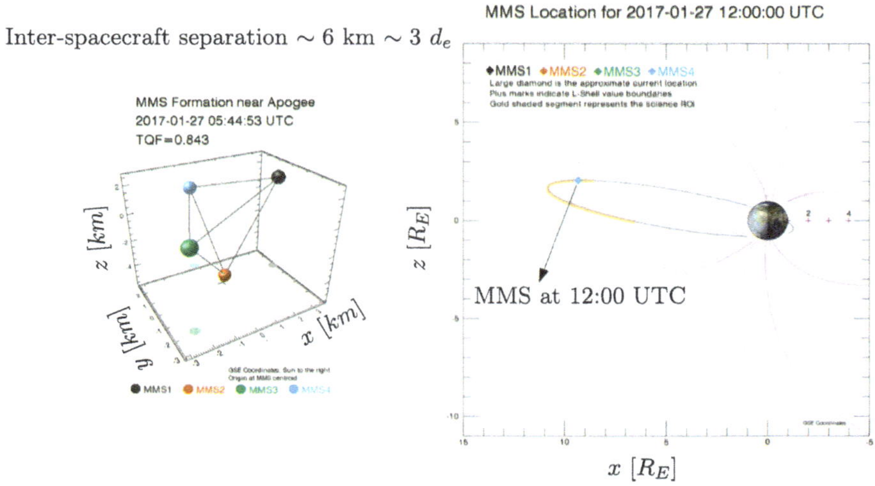

Fig. 7.1 (left) MMS tetrahedral configuration at the time of the EDR encounter; (right) MMS location relative to Earth and the average magnetopause boundary, shown in GSE, in units of Earth radii. The orbit of MMS is in black and the Region Of Interest (ROI) of the orbit is colored in yellow. The light blue diamond represents MMS at the time of the EDR encounter

4.1.4) is close to 1. This means that the results of the multi spacecraft methods of data analysis (see Sect. 4.2.2) are supposed to be rather reliable, being affected by the not-regularity of the tetrahedron.

7.2 Instrumentation

Throughout the data analysis, the burst mode data are used. In particular, the data that are used are the magnetic field data from the fluxgate magnetometer (FGM) at 128 samples/s, 3D electric field data from the axial and spin-plane probes at 8192 samples/s and particles data from the fast plasma investigation (FPI) with a time resolution of 30 ms for electrons and 150 ms for ions (see Sect. 4.1.9). Current densities are computed using single spacecraft data at the electron resolution of 30 ms (ion measurements are interpolated to the same sampling frequency of electrons).

7.3 Event Overview

This section provides an overview of the EDR event describing the magnetic field, density and velocity conditions at larger scale compared to the actual time scale of the EDR encounter. Figure 7.2 shows a 5 min interval that includes the EDR crossing marked by the yellow shaded region. Figure 7.2b–d show the MMS1 measurements of the magnetic field components, ion density, and ion velocity components in the GSE coordinate system. MMS stays mostly in the magnetospheric boundary layer, which corresponds to $B_z > 0$ and to the typical value of the density $\sim 10\,\mathrm{cm}^{-3}$ (Eastman and Hones 1979). Between 12:05:41.2 and 12:05:43.2, B_z becomes negative.

In order to understand whether the changing in sign of B_z corresponds to the crossing from magnetosphere boundary layer to the magnetosheath boundary layer we need to know the direction of the magnetic field in the magnetosheath adjacent to the magnetopause. For this reason, the WIND spacecraft (Acuña et al. 1995) data are propagated to the magnetopause using the *flat delay* method (see Sect. 4.2.1). WIND was located at (200, 100, 1) R_E (in GSE, see Fig. 7.3(top)). The solar wind velocity was rather stable, mainly directed along x and the average velocity is $V_{SW} \approx V_{SW,x} \sim 500$ km/s. It is found that the solar wind propagation delay is of 47 min. Figure 7.3(bottom) shows the magnetic field components measured by the WIND spacecraft which have been shifted using this delay to take into account propagation to the magnetopause. We can see that the magnetic field in the magnetosheath adjacent to the magnetopause was stable and directed southward, supporting the fact that when $B_z < 0$ MMS is on the magnetosheath side of the magnetopause boundary.

An ion and electron v_z jet reversals are observed at the second B_z reversal, at 12:05:43.20. The ion velocity in the z direction changes from a value of $+200$ km/s (12:05:41.0) to -150 km/s (12:05:48.0), see Fig. 7.2d. The jet reversal is observed also in the electron velocity in Fig. 7.2f and $v_{e,z}$ changes from $\sim +250$ km/s to ~ -450 km/s (the local ion Alfvén speed is $\sim$100 km/s). The high-speed ion and electron flows, the corresponding ion and electron flow reversals, as well as the B_z reversal and the low $|\mathbf{B}| \sim 3$ nT indicate that the spacecraft is in the vicinity of the reconnection region at 12:05:41.9-12:05:48.0 (yellow shaded region in Fig. 7.2a–d).

7.4 Determination of the Local Coordinate System

This section is devoted to the determination of the local coordinated system of the current sheet. Firstly, I performed a nested MVA analysis (see Sect. 4.2.1) on the current sheet crossing centered at 12:05:43.20 using MMS4 data. The length of the nested intervals on the magnetosheath side is constrained by the fact that another current sheet crossing is present at 12:05:41.20 (see Fig. 7.2e). The nested time interval showing the largest eigenvalue ratios $\lambda_L/\lambda_M \sim 40$ and $\lambda_M/\lambda_N \sim 10$ was found for the interval 12:05:41.9 - 12:05:46.9. The same interval 12:05:41.9 - 12:05:46.9 has been chosen to compute the LMN system for the other three spacecraft.

The final frame obtained with MVA is $L_{MVA} = (0.05, 0.03, 0.99)$, $M_{MVA} = (-0.30, -0.95, 0.04)$, $N_{MVA} = (0.95, -0.30, -0.04)$ is obtained by averaging the four LMN systems obtained applying MVA on each of the four spacecraft data. The nested analysis shows that the N_{MVA} direction is stable within 10° with an eigenvalue

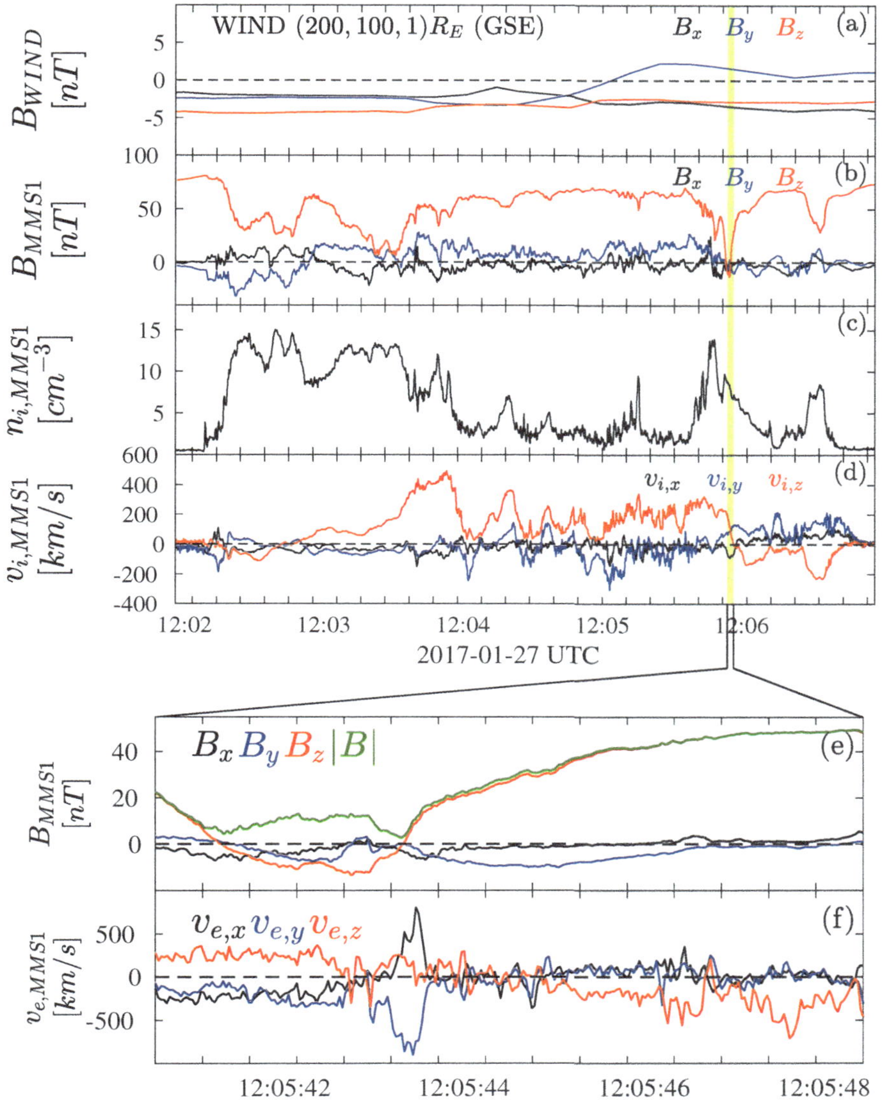

Fig. 7.2 **a** Magnetic field components as measured by WIND and propagated to the magnetopause; **b** MMS1 magnetic field components; **c** MMS1 ion density and **d** MMS1 ion velocity components; **e** Zoom-in of the MMS1 magnetic field components and strength; **f** Zoom-in of the electron velocity components. Data are shown in GSE. The yellow shaded region in panels **a**–**d** indicates the EDR crossing. © 2019 American Physical Society

ratio λ_M/λ_N always > 3. Therefore, the performed MVA analysis can be considered as quite reliable.

Then, the results from the MVA are compared with the normal direction obtained from the timing analysis (see Sect. 4.2.2). The four B_z profiles are cross-correlated by using different reference values of B_z in the interval 12:05:41.9 - 12:05:46.9 obtaining a normal direction $N_{timing} = (0.95, 0.25, 0.08)$. The results of the timing analysis are not sensitive to the chosen reference value. The normal directions obtained with the two methods are both mainly directed along the x_{GSE} direction and the angular difference between the two directions is $\sim 30°$.

In order to further validate the choice of the coordinate system, I have also computed the LMN system using the MDD method (see Sect. 4.2.2). I consider the interval 12:05:41.9 - 12:05:44.9 which encloses the current sheet. This interval is shorter than that used for the MVA and the timing analysis, but it is the longest interval where direction of the normal is stable. The mean normal direction averaged over this time interval is $N_{MDD} = (0.93, 0.16, 0.33)$. The mean values of the eigenvalue ratios are $\lambda_1/\lambda_2 \sim 100$ and $\lambda_2/\lambda_3 \sim 600$. Since $\lambda_1 \ll \lambda_2, \lambda_3$ the structure can be considered as 1D (see Sect. 4.2.2). The angle between N_{MDD} and N_{MVA} is $\sim 30°$. A similar analysis has been performed by Rezeau et al. (2018, see Table 1) on this set of data and their findings are consistent with the results presented here.

An additional rotation of 17° around the N_{MVA} direction is added to guarantee the consistency of B_M (measured by FGM) and J_L (measured by FPI) measurements within the diffusion region with the Hall currents and magnetic field pattern.

Indeed, during the current sheet crossing the MMS spacecraft are initially within the IDR (as it is discussed in Sect. 7.5) so the current densities and the magnetic field signatures are expected to be consistent with the Hall pattern (see Sect. 2.4). However, when using the LMN system obtained with MVA, there is not a full agreement between J_L and B_M. Therefore, in order to obtain consistency between J_L and B_M, I have rotated the coordinate system keeping the N direction fixed and rotating the L and M direction of 17°. Considering the previous analysis presented in this section, the angle of rotation falls within the uncertainty of the definition of the N direction. Details about this rotation are presented in Appendix 7.A to this chapter.

In conclusion, the adopted local coordinate system LMN is composed by, in GSE coordinates, $L = (-0.039, -0.252, 0.967)$, which is close to the south-north direction, $M = (-0.301, -0.921, 0.252)$, which is approximately the east-west direction and $N = (0.954, -0.300, -0.040)$, which is approximately parallel to the Earth-Sun direction. The LM plane represents the current sheet plane, where M is the direction parallel to the current, and N is perpendicular to the current sheet. In the following, data are shown in the local current sheet coordinate system LMN.

7.5 Electron Diffusion Region Signatures

The approximate trajectory of the spacecraft through the reconnection region is shown in Fig. 7.4.

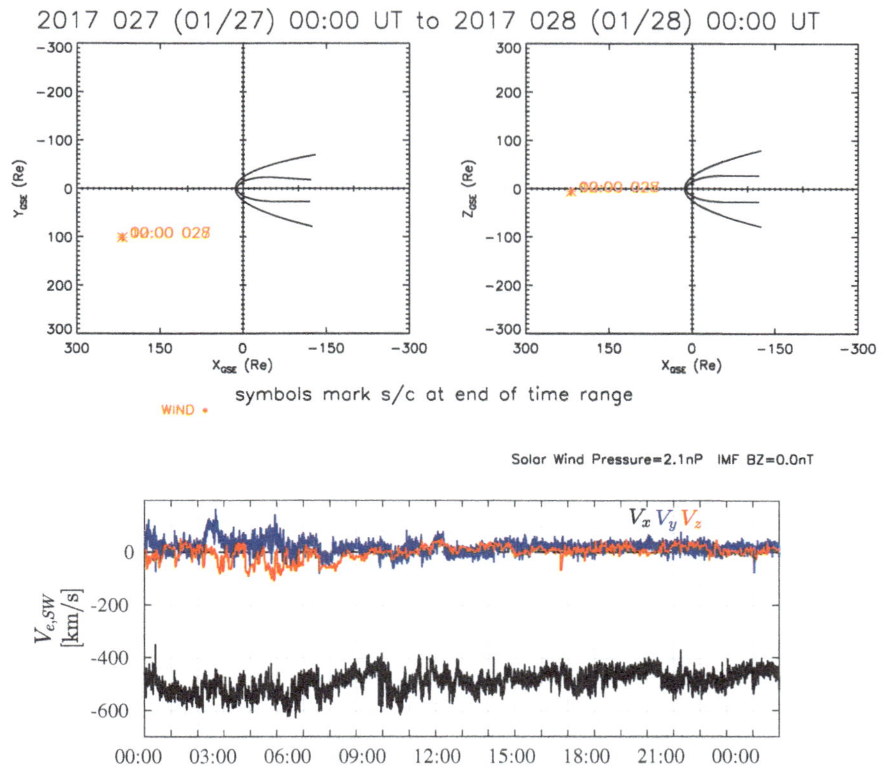

Fig. 7.3 (top) Location of the WIND spacecraft during the day of the EDR encounter (27/01/2017) in the xy and xz plane (in GSE coordinates); (bottom) components of the solar wind velocity measured by WIND on the day of the EDR encounter

In the interval shown in Fig. 7.4 (12:05:41.9 - 12:05:44.0), ions are not magnetized (see Fig. 7.6d) and B_M (Fig. 7.4b) corresponds to the out-of-plane Hall field with a distorted quadrupolar pattern, as expected for asymmetric reconnection with a weak guide field (see Sect. 3.2), with $B_M > 0$ ($B_M < 0$) on the magnetosheath side of the boundary, northern (southern) of the reconnection site. These observations indicate that the spacecraft is located in the ion diffusion region. The guide field is estimated to be less than 10% of $|\mathbf{B}|$ according to the averaged value of B_M among the spacecraft in the center of the current sheet (B_L inversion).

In interval AB (12:05:41.900 - 12:05:42.456, Fig. 7.4), all four probes observe roughly constant values of $B_L < 0$ yet showing differences of several nT despite the small inter-spacecraft separation, indicating that the current sheet is thin. A large parallel current ($J_L < 0$ in Fig. 7.4d and Hall magnetic field $B_M > 0$ (Fig. 7.4b) indicate that MMS is close to the current sheet on the magnetosheath side of the boundary, north of the reconnection site. The probes are rather close to the center

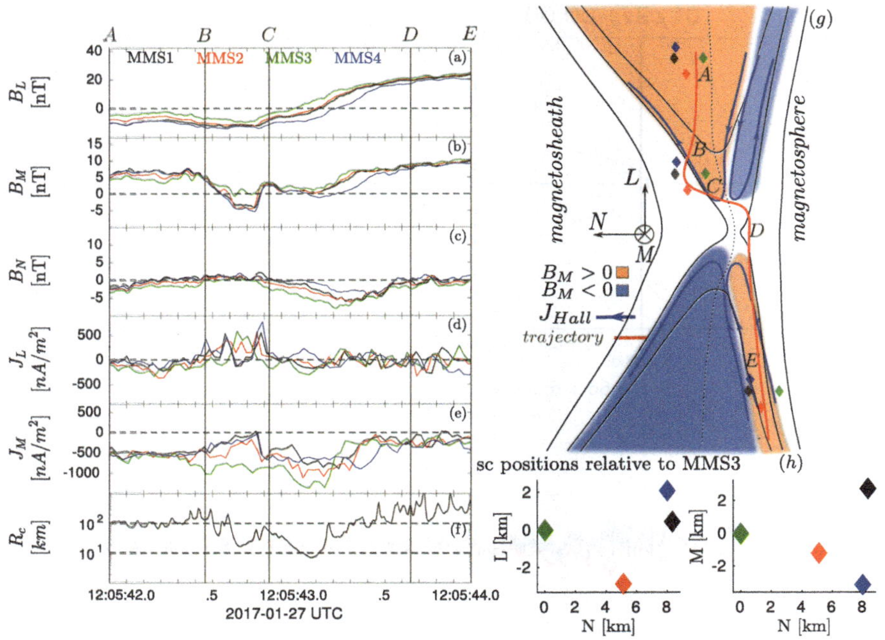

Fig. 7.4 Four spacecraft measurements of **a** B_L; **b** B_M; **c** B_N; **d** J_L; **e** J_M. **f** Curvature radius of the magnetic field lines; **g** cartoon of the encounter. The red line represent the trajectory of the barycenter of MMS constellation. Since the velocity of the magnetopause is much larger than the spacecraft velocities, the MMS path shown is produced entirely by the motion of the magnetopause in the LN plane. The three tetrahedra represent MMS location at different times along the trajectory; **h** Projection of the MMS tetrahedron in the LN and in the MN plane. © 2019 American Physical Society

of the current sheet, as indicated by the large $J_M \sim 500$ nA/m^2 and small B_L. According to the B_L difference among the probes, MMS3 is the closest to the center of the current sheet (see the tetrahedron close to location A in Fig. 7.4g) while MMS4 and MMS1 are further away. In this interval, the trajectory of MMS[3] is tangential to the magnetopause, therefore differences among the spacecraft observations have to be considered as spatial.

In interval BC (12:05:42.456 - 12:05:42.830), the peaks of $J_L > 0$ indicate that MMS moves closer to the magnetosheath separatrix. MMS1 and MMS4 make a brief excursion in the inflow region around 12:05:42.6, where the B_L gradient is smaller and all probes except MMS3 observe a minimum in J_M and $B_M \sim 0$.

[3]Note that the speed of the spacecraft is negligible compared to the speed of the magnetopause. The motion of the spacecraft is irrelevant and all of the perceived motion is due to the movement of the magnetopause. However, since it is simpler to visualize, this kind of studies usually refer to the motion of the spacecraft relative to the magnetopause.

At the same time MMS3, which is closer to the center of the current sheet, observes $B_M \sim 4$ nT and large J_M meaning that MMS3 is not observing the inflow region. Instead, given the rather large value of J_M, MMS3 is still measuring the current sheet. Accordingly, the location of the four spacecraft at this time is shown in Fig. 7.4g with the projection of the tetrahedron in the plane LN between the letters B and C indicating the corresponding time interval. After that, MMS1 and MMS4 cross again the magnetospheric separatrix and the constellation comes back in the Hall region where $B_M \sim 5$ nT for all the spacecraft (at 12:05:42.830).

In interval CD (12:05:42.83 - 12:05:43.65), MMS crosses the current sheet north of the reconnection site, as indicated by the fact that $B_N < 0$. By applying the timing method (see Sect. 4.2.2) to this current sheet crossing, I estimate the normal velocity of the current sheet to be about ~35 km/s and the normal direction to be $\mathbf{N}_{timing} = (0.95,\ 0.25,\ 0.08)$ (GSE). As previously noted, the normal direction estimated with the timing method is in good agreement with the normal found with the MVA method. According to the current sheet speed, MMS crosses an electron scale current sheet with a thickness of ~30 km $\sim 15\, d_e$. The current sheet corresponds to a strong value of $J_M > 1000$ nA/m^2. The strong decrease in B_N in the CD interval corresponds to the reconnected magnetic field. The curvature radius of the magnetic field lines $R_c = \mathbf{b} \cdot \nabla \mathbf{b}$ (where $\mathbf{b} = \mathbf{B}/|\mathbf{B}|$, Fig. 7.4f) decreases as well reaching its minimum of less than 10 km $\sim 5\, d_e$ at the $|\mathbf{B}|$ minimum (~3 nT). This indicates that the spacecraft is located close to the center of reconnection site at this time.

After the current sheet crossing (CD interval), MMS moves tangentially along the southern magnetospheric separatrix region observing a southward ion and electron jet $v_{i,L}, v_{e,L} < 0$ (corresponding to $v_{i,z}$ and $v_{e,z}$ in Fig. 7.2d and f).

7.5.1 *Computing the Distance Spacecraft-Reconnection Site with the FOTE Method*

The four spacecraft analysis of the current sheet crossing revealed that MMS is close to the reconnection site at the time of the B_L reversal (at 12:05:43.20). In order to quantify the distance between MMS and the reconnection site I apply the First Order Taylor Expansion (FOTE) method to the current sheet crossing. Since the magnetic field strength reaches a minimum of 3 nT and the guide field is small for this event, I expect that MMS is close to a magnetic null point.

The FOTE method has been discussed in Sect. 4.2.2. In this section, I just remind that the method lays on the hypothesis that the magnetic field can be linearized near the position of the null so that the distance $d\mathbf{r} = \mathbf{r} - \mathbf{r}_{null}$ between the null point and the spacecraft is given by

$$d\mathbf{r} \approx (\nabla \mathbf{B})^{-1}\mathbf{B}. \tag{7.1}$$

In this way, a null is always found. Depending on the case of study, a threshold on the magnitude of $d\mathbf{r}$ has to be imposed in order to eliminate the non-physical nulls.

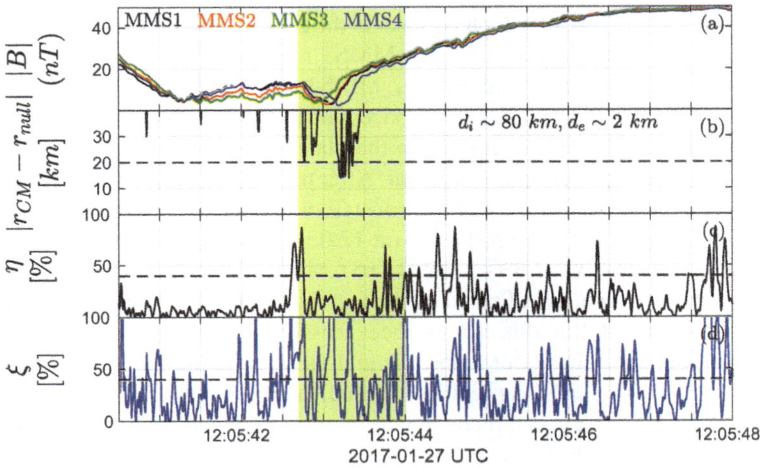

Fig. 7.5 **a** Four spacecraft magnetic field strength; **b** $|\mathbf{r}_{CM} - \mathbf{r}_{null}|$ where $\mathbf{r}_{CM}$ is the position of the center of mass of MMS tetrahedron and $\mathbf{r}_{null}$ is the position of the magnetic null point; **c** parameter $\eta = \frac{|\nabla \cdot \mathbf{B}|}{|\nabla \times \mathbf{B}|}$; **d** parameter $\xi = |\frac{(\lambda_1+\lambda_2+\lambda_3)}{\lambda_{MAX}}|$. The yellow shaded region include the interval of the EDR encounter and it corresponds to the time interval of Fig. 7.7

Since this is an event which mainly involves the electron scale, I will consider a null as physical only if the null-spacecraft distance is $< \frac{1}{4}\, d_i \sim 20\,\text{km}$.

Figure 7.5a shows the magnetic field strength of the four spacecraft in the same time interval as Fig. 7.2e–f. We can see that there are two magnetic field minima, corresponding to the two current sheet crossing discussed in Fig. 7.2. The distance between the magnetic null and the center of mass of the MMS tetrahedron is shown in Fig. 7.5b. We can clearly see that MMS is approaching the magnetic null and the minimum distance of 12 km $\sim 6\ d_e$ is reached at the magnetic minimum at 12:05:43.2, when MMS is encountering the EDR (yellow shaded area in Fig. 7.5).

The η and ξ parameters—which indicate the accuracy of the results of the method (see Sect. 4.2.2)—are both under the threshold of 40% in this interval Fig. 7.5c–d. Results with η and ξ below this threshold has been considered as reliable (Fu et al. 2015). Note that ξ is less stable than η indicating that ξ is more sensitive to the fluctuations of the magnetic field. However, ξ is also below the threshold of 40% in the interval 12:05:43.12 - 12:05:43.26, where the minimum of $d\mathbf{r}$ is reached.

As discussed in Sect. 4.2.2, the FOTE method is also able to identify the null type and this allows to characterize the EDR in terms of a quasi-2D or 3D region. However, magnetic field fluctuations can prevent the correct identification of the null type. Following the analysis by Eriksson et al. (2015), I compute the maximum value of magnetic field fluctuations δB_{MAX} above which the null type is not well defined. In this case of study, $\delta B_{MAX} \sim 0.4$ nT is comparable with the measured magnetic field fluctuations $\delta B \sim 0.7$ nT computed in the interval 12:05:43.20 and 12:05:43.25 meaning that the null type can not be identified.

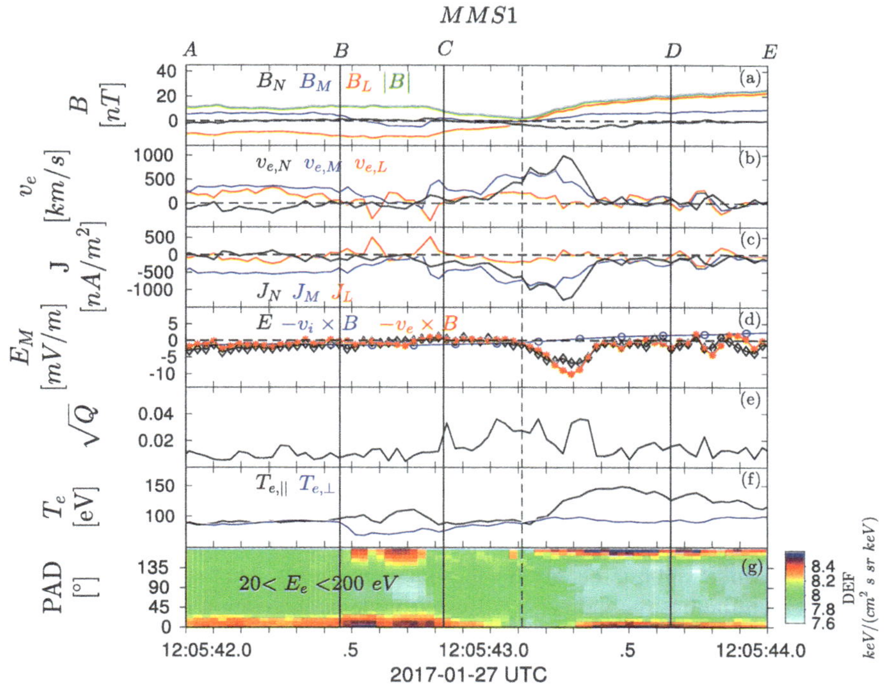

Fig. 7.6 **a** Magnetic field components and strength; **b** electron velocity components; **c** current density components; **d** M component of electric field (30 ms resolution), $(\mathbf{v}_e \times \mathbf{B})_M$ (30 ms resolution), $(\mathbf{v}_i \times \mathbf{B})_M$ (150 ms resolution); **e** agyrotropy parameter $\sqrt{Q}$; **f** parallel and perpendicular electron temperature; **g** electron pitch angle distribution in the energy range [20, 200] eV. The black vertical dashed-line indicates the time of the $|\mathbf{B}|$ minimum. Data from MMS1. © 2019 American Physical Society

The schematic trajectory of MMS (Fig. 7.4g), together with the results of FOTE method, indicates that the spacecraft crossed the magnetopause close to the reconnection site. Figure 7.6 shows further evidence of MMS crossing the EDR (see Sect. 6.1). During the magnetopause crossing identified by the B_L reversal (Fig. 7.6a), a large enhancement of the electron velocity shifted toward the magnetosphere is observed in M and N components, reaching 600 km/s and 1000 km/s respectively (Fig. 7.6b). These peaks are not observed in the ion velocity. Therefore, the current densities presented in Fig. 7.6c are carried by electrons and they peak between 12:05:43.200 and 12:05:43.350 reaching $\sim$1000 nA/m^2 in J_M and J_N. These values of J_M are expected for a current sheet at the electron scales and similar values are reported in other EDR observations (Burch et al. 2016; Webster et al. 2018). A further confirmation of the EDR encounter is given by the demagnetization of electrons (Fig. 7.6d), which are decoupled from the magnetic field ($\mathbf{E} \neq -\mathbf{v}_e \times \mathbf{B}$) between 12:05:43.150 and 12:05:43.350.

Consistently with the trajectory in Fig. 7.4, a positive $v_{e,L} \sim 400$ km/s is observed between 12:05:42.900 and 12:05:43.250 and $v_{e,L} \ll v_{A,e} \sim 4000$ km/s, the electron Alfvén speed. This indicates that MMS is crossing the inner EDR, where the electron jet has not developed yet (Karimabadi et al. 2007). Agyrotropy $\sqrt{Q}$ (see Appendix B) exhibits an enhancement in correspondence of the B_L reversal (Fig. 7.6e). The agyrotropy parameter $\sqrt{Q}$ can have non negligible values also far from the EDR, specifically along the magnetospheric separatrix (Lapenta et al. 2017) [e.g. Fig. 3], (Shay et al. 2016). Yet, in the present case, the agyrotropy increase is observed by all four MMS probes between 12:05:42.6 and 12:05:43.5 and for the majority of this interval (12:05:42.6 - 12:05:43.2) MMS is in the magnetosheath ($B_L < 0$).

The electron temperature increase is shifted towards the magnetosphere and mainly seen in the direction parallel to the magnetic field (Egedal et al. 2011; Shay et al. 2016) ($\Delta T_{e,||} \sim 50$ eV and $\Delta T_{e,\perp} \sim 25$ eV through the crossing) while at the $|\mathbf{B}|$ minimum $T_{e,||} \sim T_{e,\perp}$ (Fig. 7.6f). The same behavior is shown consistently by the electron Pitch Angle Distribution (PAD) (Fig. 7.6g). Furthermore, between 12:05:42.760 and 12:05:42.980 a low energy electron population parallel to **B** propagates toward the $|\mathbf{B}|$ minimum. At the $|\mathbf{B}|$ minimum (12:05:42.980 - 12:05:43.150) this beam is no longer observed and the PAD looks isotropic while the distribution functions exhibit oblique beams (to the magnetic field). This signature has been recently identified as the indication of electron demagnetization (see Sect. 6.1 and (Egedal et al. 2018)). In addition, the strong fluctuations in the electric field data observed in correspondence of the $|\mathbf{B}|$ minimum (Fig. 7.7e–f) suggest that high frequency waves may be present. All these EDR encounter signatures are shown using MMS1 data and they were observed overall by all probes, albeit with some differences which are significant and will be discussed below.

7.6 Electron-Scale Structuring of the EDR

Figure 7.7 show the four-spacecraft analysis of the EDR encounter. Figure 7.7a and b show respectively B_L measured by each spacecraft and the shifted B_L obtained via the timing method (see Sect. 4.2.2). The time lag between B_L components measured by MMS1 and MMS2-3-4 respectively are $\Delta t = (\Delta t_{12}, \Delta t_{13}, \Delta t_{14}) = (0.024 \text{ s}, 0.114 \text{ s}, -0.113 \text{ s})$. In order to facilitate the comparison among observations by different spacecraft, the same shift is applied to Fig. 7.7c–i.

All the probes observe a large J_M consistent with the current sheet crossing. However, while J_M reaches 1200 nA/m^2 for MMS3, its value is lower (~800 nA/m^2) for the other probes. The difference in the current density observations by different MMS probes is larger than the FPI measurement error, which is ~20% (Pollock et al. 2016). Therefore, the current densities in the EDR are not homogeneous on the scale of a few d_e, which corresponds to the spacecraft separation. To summarize, we may say that at large ion scales the current densities are homogeneous, while by looking at the electron scale we are able to observe fine structures that may be due to the

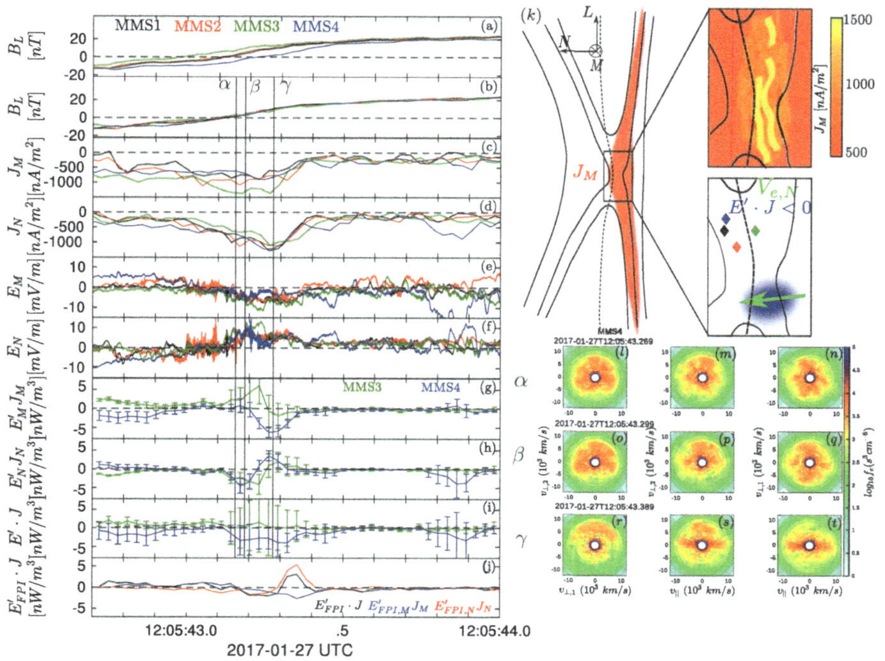

Fig. 7.7 Four spacecraft **a** B_L; **b** Time-shifted B_L. **c** Time-shifted J_M; **d** Time-shifted J_N; **e** Time-shifted E_M (8192 samples/s); **f** Time-shifted E_N (8192 samples/s); **g** Time-shifted $E'_M J_M$; **h** Time-shifted $E'_N J_N$; **i** Time-shifted $\mathbf{E}' \cdot \mathbf{J}$; **j** $\mathbf{E}'_{FPI} \cdot \mathbf{J}$, $E'_{FPI,M} J_M$, $E'_{FPI,N} J_N$. The α, β and γ lines correspond to the times of the α, β and γ distribution functions in panels **l–t** shifted accordingly to the timing method. **k** Cartoon of J_M and of the energy conversion and $v_{e,N}$ flow. **l–t** Electron distributions by MMS4 projected on $(\mathbf{v}_{\perp,1}, \mathbf{v}_{\perp,2})$, $(\mathbf{v}_{\|}, \mathbf{v}_{\perp,2})$ and $(\mathbf{v}_{\|}, \mathbf{v}_{\perp,1})$ planes at three different times t_α = 12:05:43.269, t_β = 12:05:43.299, t_γ = 12:05:43.389. © 2019 American Physical Society

filamentation of the current sheet. This idea is represented in the upper right frame of Fig. 7.7k.

The electric field E_M (Fig. 7.7e) and E_N (Fig. 7.7f) maintain the same sign during the EDR crossing. E_M and E_N are comparable and they both reach 10 mV/m. This differs from what is expected in the case of laminar and steady two-dimensional reconnection, where close to the reconnection site E_M represent the reconnection electric field and it is typically much smaller than the Hall electric field E_N. Figure 7.7d shows that a large peak of $J_N \sim -1000$ nA/m^2 is seen by all the spacecraft. Such a large $J_N < 0$ corresponds to a large $v_{e,N}$ directed toward the magnetosheath. Note that this J_N behavior is not typically observed close to the reconnection site in two-dimensional PIC simulations (Pritchett 2008; Shay et al. 2016) and observations Burch et al. (2016). Since the $v_{e,N} > 0$ region is observed by all spacecraft, the minimal width of this $v_{e,N} > 0$ *channel* has to be comparable to the spacecraft separation. In particular, in the LN plane, the minimal width of the $v_{e,N} > 0$ region is 4 km $\sim 2\, d_e$ in the L direction and of 8 km $\sim 4\, d_e$ in the N direction.

The strong J_N deeply affects the energy conversion pattern since $E'_N J_N$ (Fig. 7.7h) becomes comparable to $E'_M J_M$ (Fig. 7.7g). If we consider the maximum error associated to each quantity (with $\delta E = 20\%|E|$, $\delta B = 0.5$ nT and an error of $\sim$10% for density and velocity) we find that $E'_M J_M$ has a positive peak for MMS3 while for MMS4 $E'_M J_M$ shows a bipolar signature that is beyond the errors (Fig. 7.7g). In Fig. 7.7g–i only data from MMS3 and MMS4 are shown since they exhibit the clearest differences between spacecraft. All four probes quantities and associated errors are shown in Appendix 7.A.

The energy conversion errors are comparable to the measured quantities for all the spacecraft. However, on MMS4 errors are smaller so that an unambiguous value for the total $\mathbf{E}' \cdot \mathbf{J} \sim E'_M J_M + E'_N J_N$ ($E'_L J_L \ll E'_M J_M$, $E'_N J_N$) is obtained on MMS4. In particular, on MMS4 $\mathbf{E}' \cdot \mathbf{J} < 0$ (Fig. 7.7i), showing negative energy transfer between fields and particles. This indicates that energy is locally converted from the particles to the field, the opposite of the standard behavior during reconnection. The local energy conversion is negative especially because of the contribution of $E'_N J_N$ which is non-negligible because J_N is larger than usually observed. The correlation between the channel with $v_{e,N} > 0$ and the $\mathbf{E}' \cdot \mathbf{J} < 0$ is represented in the sketch in the bottom right panel of Fig. 7.7k.

Since MMS4 is the only spacecraft that provides a value of the energy conversion $\mathbf{E}' \cdot \mathbf{J}$ beyond the errors, I have also computed the electric field using Ohm's law

$$\mathbf{E}'_{FPI} = -\frac{\nabla \overline{\mathbf{P}}_e}{ne} + \frac{m_e}{e}\mathbf{v}_e \cdot \nabla \mathbf{v}_e + \frac{m_e}{e}\frac{\partial}{\partial t}\mathbf{v}_e \qquad (7.2)$$

Here, $\overline{\mathbf{P}}_e$ is the electron pressure tensor and the subscript FPI indicates that $\mathbf{E}'_{FPI}$ is obtained by using measurements from FPI instrument only. $\nabla \overline{\mathbf{P}}_e$ is calculated using four spacecraft measurements and the full pressure tensor (see Sect. 4.2.2) so it is an average over the spacecraft tetrahedron. Note that the errors on particles data provided by FPI are smaller than the electric field errors.

It is found that, since the contribution of the inertia term is negligible (not shown, the maximum value of the inertia term in the considered time interval is $\sim$0.1 mV/m), a good proxy for the electric field is $\mathbf{E}'_{FPI} = -\nabla \overline{\mathbf{P}}_e / ne$. The quantities $E'_{FPI,N} J_N$ (Fig. 7.7j) and $E'_{FPI,M} J_M$ (Fig. 7.7j), exhibit bipolar signatures, as the total energy conversion $\mathbf{E}'_{FPI} \cdot \mathbf{J}$ (Fig. 7.7j). Yet, it should be noted that $\mathbf{E}'_{FPI}$ is a four-spacecraft measurement averaged over the tetrahedron and one should be careful when comparing it to single spacecraft observations especially if, as in this case, significant differences are seen among probes observations. For consistency, $\mathbf{J}$ is the current density which is also averaged over the tetrahedron in this case. After a careful evaluation of all error sources, I conclude that the discrepancy between the *punctual* (as given by MMS4) and the *averaged* energy conversion (given by $\mathbf{E}'_{FPI} \cdot \mathbf{J}$) is not an instrumental effect and indicates that energy conversion is not homogeneous over the tetrahedron and that energy conversion is patchy over scales of the order of few d_e.

The evolution of the electron distribution functions (DFs) measured by MMS4 in the EDR is shown in Fig. 7.7l–t. The projection of the electron DFs are made in the three perpendicular planes $(\mathbf{v}_{\perp,1}, \mathbf{v}_{\perp,2})$, $(\mathbf{v}_{||}, \mathbf{v}_{\perp,2})$ and $(\mathbf{v}_{||}, \mathbf{v}_{\perp,1})$ where $\mathbf{v}_{\perp,1} = \mathbf{v} \times \mathbf{b}$, $\mathbf{v}_{\perp,2} = \mathbf{b} \times (\mathbf{v} \times \mathbf{b})$ and $\mathbf{v}_{||} = \mathbf{b}$ ($\mathbf{v} = \mathbf{v}_e/|\mathbf{v}_e|$ and $\mathbf{b} = \mathbf{B}/|\mathbf{B}|$) and at the three times indicated by the vertical black lines in Fig. 7.7b–i. The times are shifted according to the delays among spacecraft obtained with the timing method. These times correspond to regions where $E'_M J_M$ is positive (DFs indicated with α, Fig. 7.7l–n), negative (DFs indicated with β, Fig. 7.7r–t) and in the transition from positive to negative (DFs indicated with γ, Fig. 7.7o–q). Similar DFs are observed by all spacecraft and they last for more than one FPI measurement (with 30 ms resolution). The α DFs (Fig. 7.7l–n) have a rather complicated shape with several oblique beams. This pattern is observed around the magnetic field minimum, from 12:05:43.179 to 12:05:43.269 for MMS4. When $E'_M J_M$ changes sign, the DFs change shape (Fig. 7.7o–q) and clearly become crescent-shaped distributions in the $(\mathbf{v}_{\perp,1}, \mathbf{v}_{\perp,2})$ plane when $E'_M J_M < 0$ (Fig. 7.7r). The DFs observed during this EDR encounter are rather complex. They are not always crescent-like and they appear to be related to $E'_M J_M$. Further analysis and comparisons with simulations are needed to fully understand the dynamics of electrons in such a complex EDR.

7.7 Discussion and Conclusions

This chapter presents observations of an Electron Diffusion Region (EDR) encountered at the magnetopause by the MMS spacecraft with the very low inter-spacecraft separation of $\sim$3 d_e. During this electron-scale current sheet crossing the four MMS spacecraft observe typical EDR signatures (see Sect. 6.1) suggesting that MMS crossed the magnetopause in close proximity to a X-line. These signatures include a large current density mainly carried by electrons (Fig. 7.6b–c), a peak of electron agyrotropy (Fig. 7.6e), demagnetization of ions and electrons (Fig. 7.6d–g), increased electron temperature anisotropy with $T_{e,||} > T_{e,\perp}$ (Fig. 7.6f), crescent-shaped electron distribution functions (Fig. 7.7o–r). Furthermore, the electron jet has not fully developed ($v_{A,i} < v_{e,L} < v_{A,e}$) indicating that MMS is within the inner EDR (Karimabadi et al. 2007).

Another observed inner EDR signature is the fact that low energy field-aligned electron beams directed towards the X-line become oblique in close proximity to the center of the EDR (Fig. 7.6g). This behaviour indicates electron demagnetization (see Sect. 6.1). Indeed, 2D kinetic simulations (Egedal et al. 2018) showed that the transition from the field-aligned distribution to the one with oblique beams takes place where the magnetic field is sharply changing direction and has the smallest magnitude, leading to the electron decoupling from the magnetic field.

In the presented event, all four MMS probes observed the EDR signatures. The multi-spacecraft analysis of the EDR revealed that the current density J_M is spatially inhomogeneous at electron scales (Fig. 7.7c and k). Previously reported EDR encounters (see Sect. 6.1, e.g. (Burch et al. 2016; Chen et al. 2017)) do not point out differences among spacecraft in the current density observations. It might be possible that the inter-spacecraft separation in these EDR encounters was not small enough to have all the spacecraft within the EDR and to resolve the electron scale inhomogeneities. However, it might be also possible that the EDR becomes structured at electron scales under particular conditions, e.g. related to the guide field value or the reconnection inflow conditions. Indeed, similar inhomogeneities have been seen in high-resolution PIC simulations (Jara-Almonte et al. 2014) where the current density is found to be structured at electron scale and below.

Strikingly, in the center of the reconnection site, the current density in the direction normal to the current sheet, J_N, is observed to have almost the same magnitude as the out-of-plane J_M current density (Fig. 7.7c–d). In addition, electrons are observed to move from the magnetosphere to the magnetosheath side of the magnetopause, corresponding to $J_N < 0$. This behaviour differs from the typical observations of the electron motion close to the reconnection site (Burch et al. 2016) as well as predictions by 2D PIC simulations as e.g. (Pritchett 2008; Shay et al. 2016). However, the observations presented in this chapter are consistent with recent PIC simulations with low numerical noise (Egedal et al. 2018; Swisdak et al. 2018) in which electrons move downstream along the magnetospheric separatrix performing oscillations of decaying amplitude in the N direction. These oscillations are shown in Fig. 7.8 which are adapted from Swisdak et al. (2018) and Egedal et al. (2018). The velocity oscillations observed in simulations (Egedal et al. 2018; Swisdak et al. 2018) are composed by alternating regions, or channels, of positive and negative $v_{e,N}$ that can be seen in Fig. 7.8b–c. In the EDR encounter presented in this chapter, such oscillations are not observed (Fig. 7.7d), which might indicate that all the spacecraft were measuring the same channel with $v_{e,N} > 0$. Accordingly, the channel's width has to be comparable to or larger than the inter-spacecraft separation of $\sim 3\ d_e$.

Another characteristic of the presented EDR is the similarity in magnitude of the electric field E_M and E_N components. This has been identified as one of the signatures of inhomogeneous current layer "disrupted" by turbulence in three-dimensional simulations (Price et al. 2016). Accordingly, our observations support the picture of the EDR as the site of strong spatial gradients and inhomogeneities.

The energy conversion $\mathbf{E}' \cdot \mathbf{J}$ (Fig. 7.7i) is highly affected by the $J_N \sim J_M$ and $E_M \sim E_N$ behavior since the two terms $E'_M J_M$ and $E'_N J_N$ become comparable (Fig. 7.7g–h). In other EDR encounters by MMS (Burch et al. 2016, 2018), $\mathbf{E}' \cdot \mathbf{J} \sim E'_M J_M$ since J_N is usually negligible in comparison to J_M. For the EDR presented here, the multi-spacecraft analysis revealed that energy conversion $\mathbf{E}' \cdot \mathbf{J}$ is spatially inhomogeneous at electron scales. I have also shown that the quantitative evaluation of energy conversion is affected by the experimental errors (Fig. 7.7g–i). However, the comparison of the single spacecraft measurements from different spacecraft (Fig. 7.7g–i) and the measurements averaged over the tetrahedron (Fig. 7.7j) both support the qualitative picture in which $\mathbf{E}' \cdot \mathbf{J}$ is patchy and changing sign in the vicinity of the reconnection site. This implies that the EDR comprises of regions where energy is transferred from the field to the plasma and regions with the opposite energy transition, which is unexpected during reconnection. A negative energy conversion was also observed in the outer EDR (Hwang et al. 2017).

Electron-scale variations of $\mathbf{E}' \cdot \mathbf{J}$ in the EDR have been recently observed (Burch et al. 2018). However, in Burch et al. (2018) these variations are oscillations of $\mathbf{E}' \cdot \mathbf{J}$ which are the consequence of the oscillatory electric field pattern that shows signatures of a standing wave. This differs from the $\mathbf{E}' \cdot \mathbf{J}$ behavior reported in our study where no such oscillatory behavior of the electric field is observed and the patchy energy conversion is consistent to spatial inhomogeneities due to electron scale structuring.

The origin of the patchy energy conversion appears to be connected to the large $v_{e,N} \sim v_{e,M}$ directed from the magnetosphere to magnetosheath that has never been observed before. This behaviour is consistent with simulations results shown in Fig. 7.8 where correlation has been found between the oscillations of the electron flow in the N direction and the oscillatory and patchy behaviour of the energy conversion.

Further observational cases as well as 3D PIC simulations with higher resolution and lower noise or full Vlasov simulations are required to understand which conditions may lead to the structuring of the EDR and how this patchy structure may affect the electron energization. These observations can be an indication of what might be observed in the EDR in the magnetotail, where highly detailed observation are available since the inter-spacecraft separation of MMS is of the order of 1 d_e.

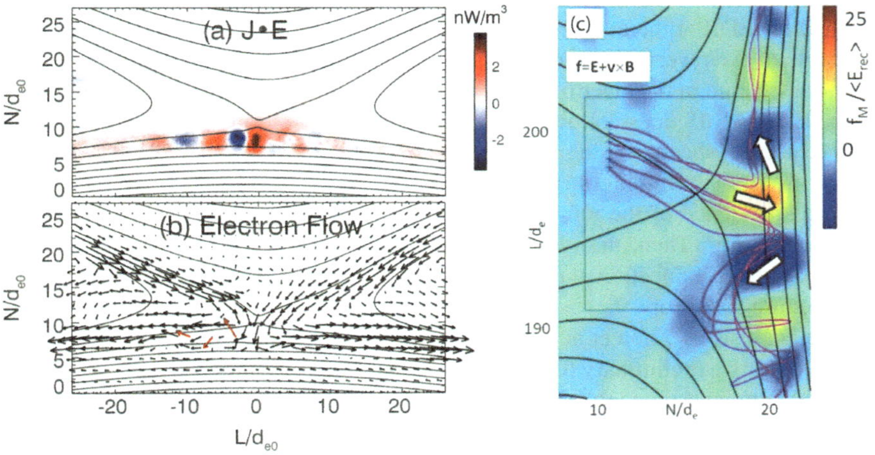

Fig. 7.8 **a** Contour plot of the energy conversion $\mathbf{E} \cdot \mathbf{J} \approx \mathbf{E}' \cdot \mathbf{J}$. Adapted from Swisdak et al. (2018); **b** Electron flow. The arrows highlighted in red indicate the oscillations of $v_{e,N}$. Adapted from Swisdak et al. (2018). © 2018 by the American Geophysical Union; **c** Contours of $f_M/\langle E_{rec}\rangle = (\mathbf{E} + \mathbf{v} \times \mathbf{B})_M/\langle E_{rec}\rangle$ where $\langle E_{rec}\rangle$ is the is the time average of the reconnection electric field for the period where reconnection is ongoing. The magenta lines are electrons trajectories. The three white arrows show the direction of the electron flow. Reprinted figure with permission from J. Egedal et al., Phys. Rev. Lett. **120**, 055101 (2018) Egedal et al. (2018). © 2018 by the American Physical Society

7.8 Future Work

Throughout this chapter, I have reported MMS observations of an EDR which is structured at electron scales. The encounter took place during the Phase 1 of the MMS mission, when MMS was sampling the dayside magnetosphere. During Phase 2 (see Sect. 4.1.3) MMS sampled the magnetotail. Hence, one of the immediate outlook of this observational study is to perform a similar analysis during an EDR encounter at the magnetotail in order to understand if the EDR can be inhomogeneous and exhibiting patchy energy conversion also in the magnetotail. In particular, the question is whether the EDR can be inhomogeneous in different environment or the structuring is specific to the magnetopause. And if this is the case, why are these inhomogeneities arising only at the magnetopause?

At the moment, the study of the EDR in the magnetotail is still preliminary. In particular, I have devoted some time to the selection of an EDR encounter event. Finally, I selected an EDR encounter taking place on the 10th of August 2017 at around 12:18:32 UTC. At that time, MMS was located at $(-15.2,\ 4.6,\ 3.0)\ R_E$ in Geocentric Solar Magnetospheric (GSM) coordinates. The spacecraft separation, which is about 20 km, is at electron scales since it corresponds to $\sim 2\ d_e$ (Fig. 7.9).

This event has been reported by Zhou et al. (2019) as an example of EDR encounter with a larger guide field (13% of the reconnecting field) than the one analyzed by Torbert et al. (2018), where the guide field was ~4% of the reconnecting magnetic field. In this study, the EDR is identified by using the same criteria that has been defined for EDR at the magnetopause (see Sect. 6.1). The main conclusion is that a guide field, even weak, can modify the electron dynamics in the EDR. In particular, even if a change in the electron energy has been observed from the upstream region to the EDR, MMS does not observe energetic electrons during the encounter.

The current sheet crossing takes place between 12:18:29 and 12:18:37 UTC. As shown in Fig. 1 in Zhou et al. (2019), the reversal of B_z (in GSM) is taking place in correspondence of the ion jet reversal in $v_{i,x}$ and of the B_x reversal, indicative of reconnection. Then, the typical signatures of the EDR are observed including strong out-of-reconnection plane current J_M, enhanced electron agyrotropy and large temperature anisotropy $T_{e,||} > T_{e,\perp}$.

The analysis by Zhou et al. (2019) is mainly a single-spacecraft study. However, by performing a preliminary four spacecraft analysis, one can see that this event may be suitable to investigate the structure of the EDR with multi-point observations and that it shows some signatures that has been already discussed for the magnetopause event.

In Fig. 7.10 the data are shown in the local coordinate system LMN which as been defined in Zhou et al. (2019). The L axis is roughly directed along the x_{GSM} direction, the M component is directed along the dawn-dusk direction pointing towards dusk and N is roughly directed along z_{GSM} and it is normal to the current sheet. Figure 7.10a shows the L component of the magnetic field of the four spacecraft which changes sign at ~12:18:32.8 indicating the middle of the current sheet. At the time of the B_L reversal, the out-of-reconnection-plane current density J_M reaches its

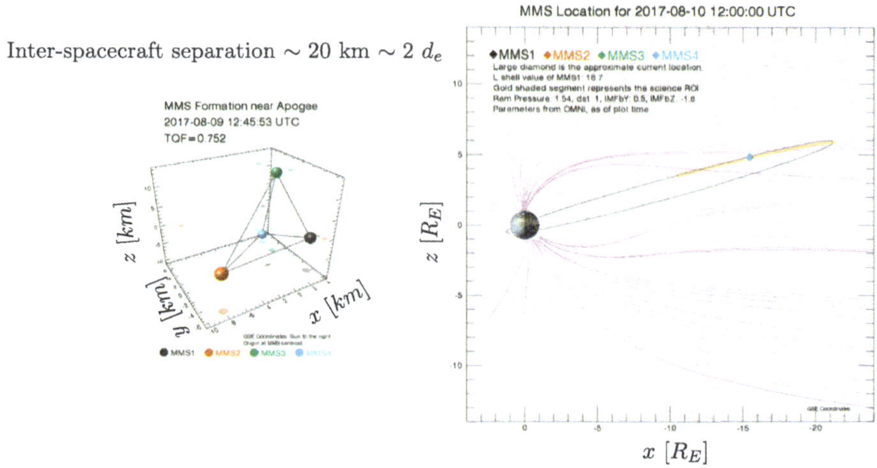

Fig. 7.9 (left) MMS tetrahedral configuration at the time of the EDR encounter; (right) MMS location relative to Earth, shown in GSE, in units of Earth radii. The magenta lines represents the magnetic field lines. The orbit of MMS is in black and the Region Of Interest (ROI) of the orbit is colored in yellow. The light blue diamond represents MMS at the time of the EDR encounter

maximum value of ~200 nA/m^2 for all the probes (Fig. 7.10b). The behaviour of J_M is rather similar for all the probes. However, MMS1 seems to observe several current peaks at 12:18:31.5 and 12:18:32.5 while the current profiles observed by the other spacecraft appear to be smoother. We need to carefully establish the trajectory across the current sheet before drawing any conclusion and this will be done in future work.

One of the main observations of the EDR magnetopause crossing that has been analyzed throughout this chapter is the enhanced $J_N \sim J_M$ in close proximity of the reconnection site. In this case, J_N is negligible compared to J_M (Fig. 7.10c).

On the other hand, similarly to the EDR encounter at the magnetopause, the electric field components E_M and E_N (and E_L) are rather similar in magnitude (Fig. 7.10d–f). Even though the spacecraft separation is ~2 d_e the electric field signal looks different on the four spacecraft. For instance, while E_M is positive for MMS1 and MMS2 in the region of enhanced current (12:18:31 – 12:18:36), it exhibit a dipolar signature on MMS3 and it is mostly negative on MMS4 (Fig. 7.10e). Differences are observed also on the other components and also on the largest component E_N. In particular, while E_N becomes clearly negative (reaching 30 mV/m) for MMS2 and MMS4 in the region with positive B_L, it fluctuates around zero for MMS1 (Fig. 7.10f). In addition, both E_M and E_N are changing sign during the crossing for MMS3 and MMS4 but while this is expected for E_N—it reflects the Hall dynamics—this behavior has to be clarified for E_M. Then, E_L reaches 10 mV/m and it stays mostly positive for MMS3 while is rather large (~5 mV/m) and negative for MMS4 (Fig. 7.10d).

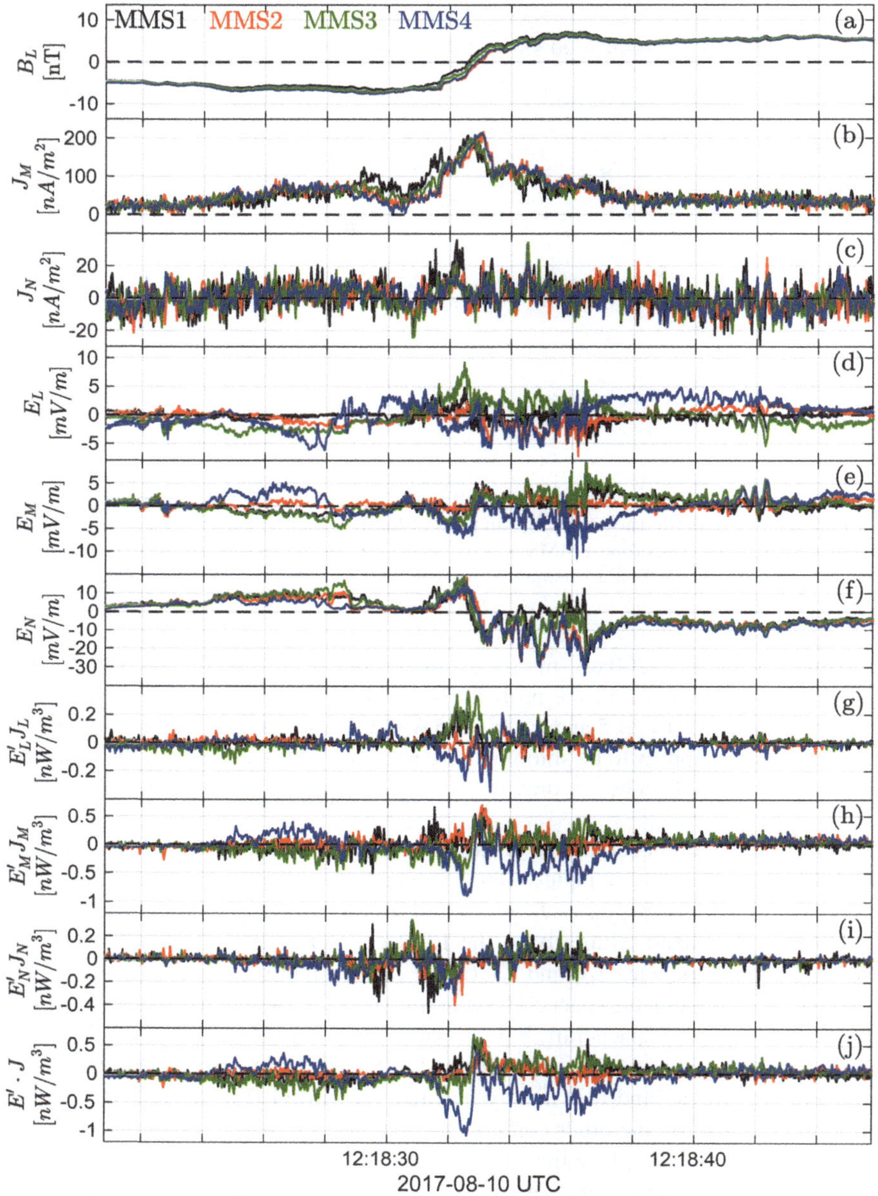

Fig. 7.10 Four spacecraft signatures of the EDR encounter in the magnetotail. Four spacecraft (**a**) B_L; (**b**) J_M; (**c**) J_N; (**d**) E_L; (**e**) E_M; (**f**) E_N; (**g**) $E'_L J_L$; (**h**) $E'_M J_M$; (**i**) $E'_N J_N$; (**j**) $\mathbf{E}' \cdot \mathbf{J}$.

These differences in the electric field observations affect the energy dissipation. $E'_L J_L$ is mostly positive for MMS1 and MMS3 while it is negative for MMS4 and MMS2 (Fig. 7.10g). The contribution of $E'_N J_N$ (Fig. 7.10i) is small compared to the others in this case, since J_N is negligible. The main contribution to energy conversion is provided by $E'_M J_M$ which—similarly to the magnetopause case—is bipolar for MMS3 while it stays mostly positive for MMS1 and MMS2 (Fig. 7.10h). $E'_M J_M$ behaviour is more complex for MMS4 but $E'_M J_M$ is mainly negative and larger that on the other probes.

The energy conversion $\mathbf{E}' \cdot \mathbf{J}$ (Fig. 7.10j) is also changing sign during the crossing on MMS3 and MMS4 indicating that there are regions in which energy is converted back from the particles to the magnetic field (Fig. 7.10j) and that the energy conversion may be patchy. This is particularly clear for MMS4 but I underline that these measurements have to be taken with care. As discussed in Sect. 7.6, the electric field measurement are sensitive and are usually affected by large errors that have to be taken into account. For the moment, I have not performed the error analysis.

In conclusion, in this section I have discussed an EDR encounter in the Earth's magnetotail which present some signatures in common (but also come differences) with the magnetopause event involving a structured EDR presented in this chapter. The preliminary analysis of this event and the comparison with the magnetopause EDR crossing brings on several questions.

1. Is energy conversion patchy within the EDR at the magnetotail? What are the processes underneath this non-uniform energy conversion?
2. Why $E'_M J_M$ exhibit a dipolar signature across the current sheet? This behavior seems rather general since it has been also observed at the magnetopause.
3. Are the differences among spacecraft observations in $\mathbf{E}' \cdot \mathbf{J}$ physical or they will not be meaningful once the uncertainties are also taken into account?
4. In this case J_N is negligible compared to the out-of-reconnection plane current J_M. At the magnetopause, the enhanced J_N has been interpreted as due to electrons exiting the EDR and performing oscillatory trajectories (see Sect. 7.7 and in particular Fig. 7.8). Should we expect to observe this behaviour also in the magnetotail? Are this oscillation present in both asymmetric and symmetric configurations or they arise only when an asymmetry is present?

All these questions will not be answered in the context of this Thesis but for sure it is worth to continue this kind of investigation. No clear conclusions can be drawn before having determined the precise trajectory of MMS relatively to the EDR and before having taken into account the relative errors of the different quantities involved in the analysis.

7.A Supplemental Material About the Determination of the Local Coordinate System

As already pointed out in Sect. 7.4, during the current sheet crossing the MMS spacecraft are initially within the IDR (as discussed in Sect. 7.5) so the current densities and the magnetic field signatures are expected to be consistent with the Hall pattern (see Sect. 2.4). However, when using the LMN system obtained with the MVA, there is not a full agreement between J_L and B_M. Here, this inconsistency is discussed and it is shown that a rotation the system of 17° around N_{MVA} is sufficient to eliminate the inconsistency without affecting other signatures.

Figure 7.11 shows B_z, B_y and J_z in the GSE coordinate system, in the LMN coordinate system obtained with MVA and in the LMN coordinate system further rotated of 17° degrees around N_{MVA} direction. Since the LMN systems are actually rather similar to GSE, Fig. 7.11 shows in the same panel the GSE components and the LMN components that are closest between each other (z and L, y and $-M$). Being the z (L) direction the maximum variance direction, the z (L) component of both

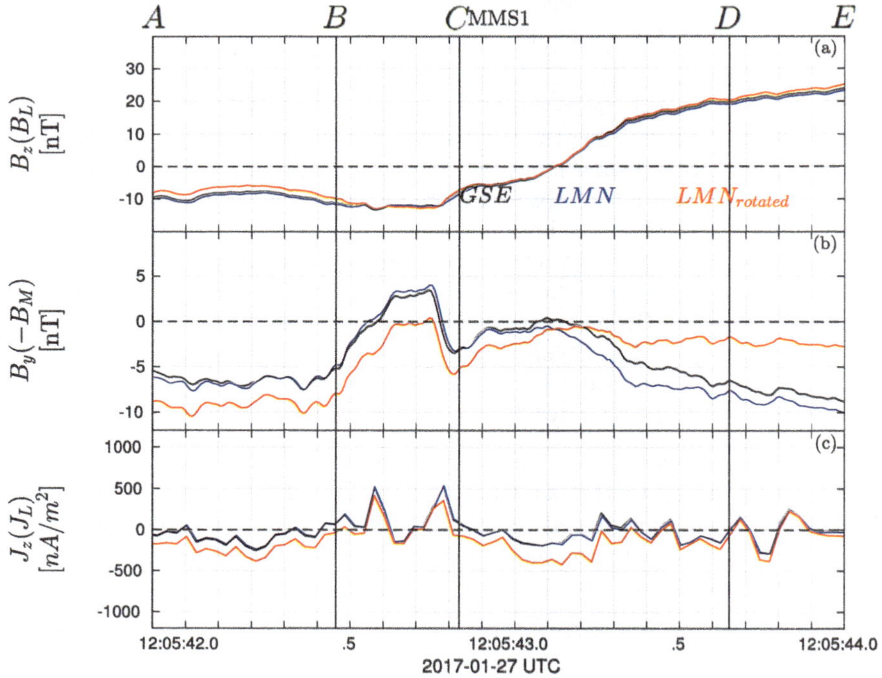

Fig. 7.11 **a** B_z and B_L magnetic field components in the GSE (black), LMN (blue) and LMN$_{rotated}$ (red) coordinate systems; **b** B_y and $-B_M$ magnetic field components in the GSE (black), LMN (blue) and LMN$_{rotated}$ (red) coordinate systems; **c** J_z and J_L current density components in the GSE (black), LMN (blue) and LMN$_{rotated}$ (red) coordinate systems;

magnetic field and current are almost not affected by the changing in the coordinate system. On the other hand, without the rotation B_M changes sign twice in interval BC.

As discussed in Sect. 7.5, in interval AB MMS is located north of the reconnection site on the magnetosheath side and $B_M > 0$ ($-B_M < 0$) there. In interval BC, B_L is still negative so MMS is still on the magnetosheath side. The positive peaks in J_L in interval BC are consistent with MMS crossing the magnetosheath separatrix north of the reconnection site. The fact that $-B_M$ becomes positive in interval BC would imply that MMS move south of the reconnection site on the magnetosheath side (B_L is still negative) and that MMS would come back again north of the reconnection site, as shown by the green line in Fig. 7.12. Indeed, the magnetosheath separatrix is observed again (second peak in J_L) and $-B_M$ becomes negative again at the end of interval BC. In such scenario, we would expect MMS to observe also the magnetosheath separatrix south of the reconnection site between the two positive peaks of J_L but the value of J_L is not consistent with this scenario (see Fig. 7.12).

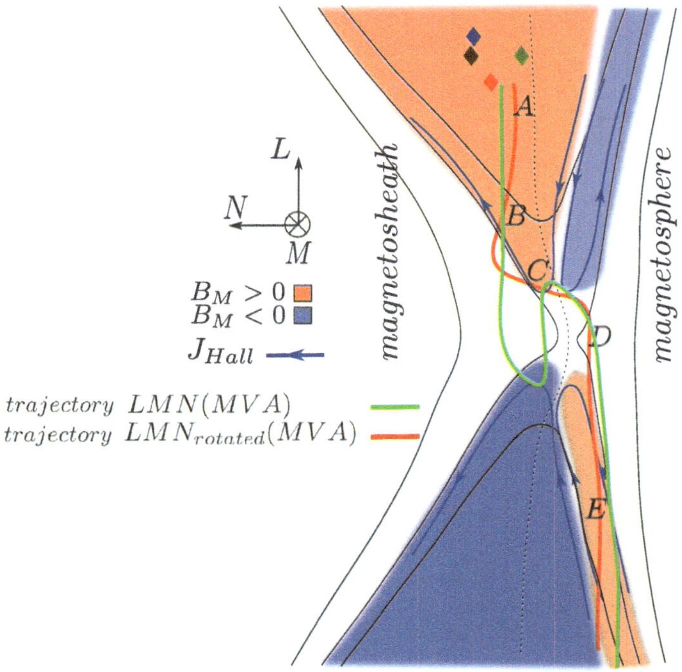

Fig. 7.12 Illustration of the magnetopause crossing as inferred from the data shown in the LMN (green line) or in the LMN$_{rotated}$ coordinate systems. The four diamonds represents the projection of MMS constellation on the LN plane

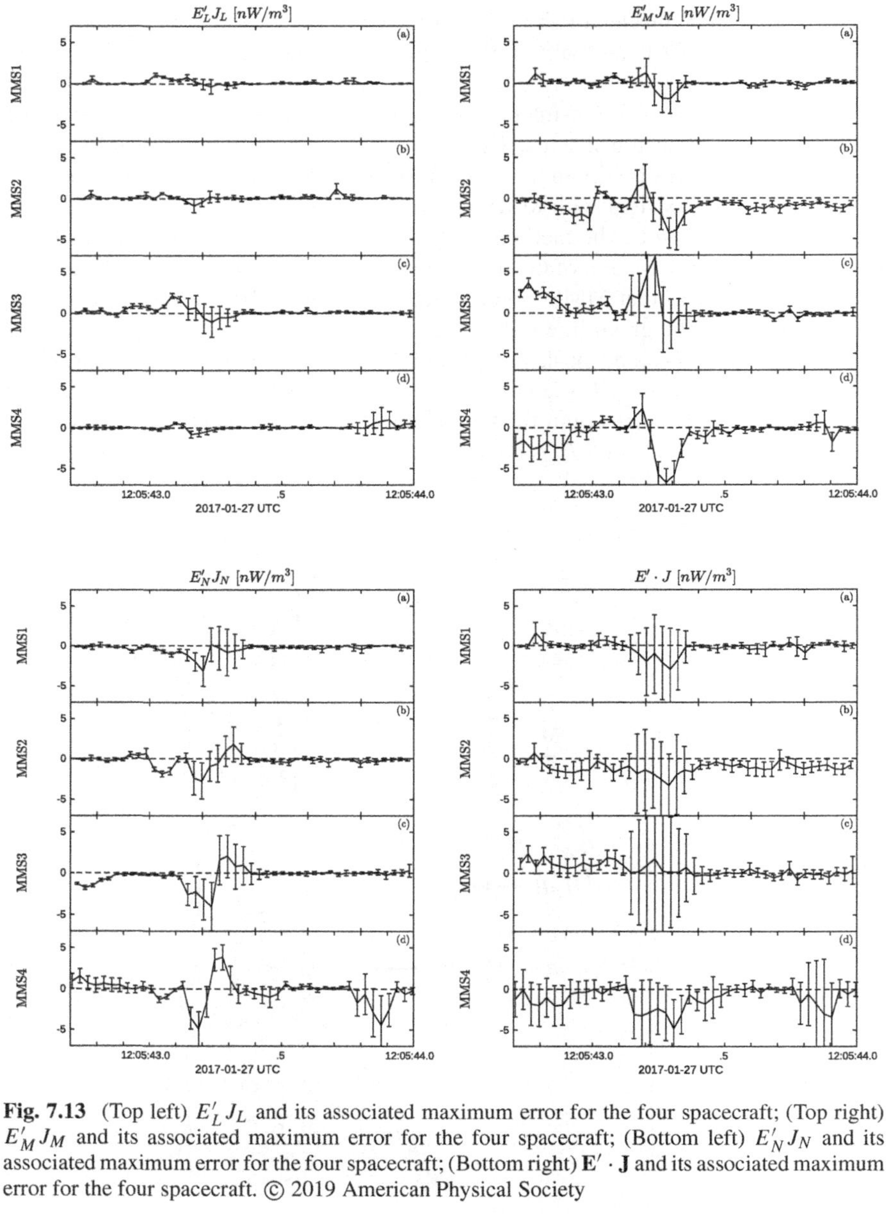

Fig. 7.13 (Top left) $E'_L J_L$ and its associated maximum error for the four spacecraft; (Top right) $E'_M J_M$ and its associated maximum error for the four spacecraft; (Bottom left) $E'_N J_N$ and its associated maximum error for the four spacecraft; (Bottom right) $\mathbf{E'} \cdot \mathbf{J}$ and its associated maximum error for the four spacecraft. © 2019 American Physical Society

Therefore, in order to obtain consistency between J_L and B_M I have rotated the coordinate system keeping the N direction fixed and rotating the L and M direction of 17°.

7.A Supplemental Material About $\mathbf{E}' \cdot \mathbf{J}$ and the Associated Error for the Four Spacecraft

This section[4] discusses the computation of the maximum errors on $\mathbf{E}' \cdot \mathbf{J}$, $E'_L J_L$, $E'_M J_M$, $E'_N J_N$. The maximum error on $E'_i J_i$ is computed as follows

$$\delta(E'_i J_i) = J_i \, \delta E'_i + E'_i \, \delta J_i \tag{7.3}$$

in which all the quantities are supposed to be not correlated and $i = L, M, N$. The error on the electric field is $\delta E_i = 20\%|E|$ when $|E| > 1$ mV/m and $\delta E_i = 0.1$ mV/m otherwise, based on statistical analysis on $\mathbf{E}$ and δE_i. The errors on the electron and ions moments are $\sim$10% and the magnetic field error is $\delta B = 0.5$ nT. The error on $\mathbf{E}' \cdot \mathbf{J}$ is $\delta(\mathbf{E}' \cdot \mathbf{J}) = \sum_{i=L,M,N} \delta(E'_i J_i)$. The behavior of the different dissipation terms is qualitatively the same for all the spacecraft, for example $E'_N J_N$ and $E'_M J_M$ are bipolar for all the probes (see Fig. 7.13). Nevertheless, the errors are comparable to the measured quantities except for MMS4, the only satellite for which the negative and positive peaks of the dissipation terms are beyond the errors.

References

M.H. Acuña, K.W. Ogilvie, D.N. Baker, S.A. Curtis, D.H. Fairfield, W.H. Mish, Space Sci. Rev. **71**, 5 (1995)
J.L. Burch et al., Geophys. Res. Lett. **45**, 1237 (2018)
J.L. Burch et al., Science **352**, aaf2939 (2016)
H. Che, J.F. Drake, M. Swisdak, Nature (London) **474**, 184 (2011)
L.-J. Chen et al., J. Geophys. Res. **122**, 5235 (2017)
G. Cozzani et al., Phys. Rev. E **99**, 043204 (2019)
W. Daughton, V. Roytershteyn, H. Karimabadi, L. Yin, B.J. Albright, B. Bergen, K.J. Bowers, Nat. Phys. **7**, 539 (2011)
T.E. Eastman E.W. Hones Jr., J. Geophys. Res. **84**, 2019 1979)
J.P. Eastwood, T.D. Phan, S.D. Bale, A. Tjulin, Phys. Rev. Lett. **102**, 035001 (2009)
J. Egedal et al., Phys. Rev. Lett. **120**, 055101 (2018)
J. Egedal, A. Le, P.L. Pritchett, W. Daughton, Phys. Plasmas **18**, 102901 (2011)
R.E. Ergun et al., Geophys. Res. Lett. **44**, 2978 (2017)
E. Eriksson, A. Vaivads, Yu.V. Khotyaintsev, V.M. Khotyayintsev, M. André, Geophys. Res. Lett. **42**, 6883 (2015)
H.S. Fu et al., Geophys. Res. Lett. **44**, 37 (2017)
H.S. Fu et al., J. Geophys. Res. **120**, 3758 (2015)

[4] Which is covering the Supplemental Material of Cozzani et al. (2019).

D.B. Graham et al., J. Geophys. Res. **122**, 517 (2017)
K.-J. Hwang et al., Geophys. Res. Lett. **44**, 2049 (2017)
J. Jara-Almonte, W. Daughton, H. Ji, Phys. Plasmas **21**, 032114 (2014)
H. Karimabadi, W. Daughton, J. Scudder, Geophys. Res. Lett. **34**, L13104 (2007)
G. Lapenta et al., J. Geophys. Res. **122**, 2024 (2017)
G. Lapenta, S. Markidis, M.V. Goldman, D.L. Newman, Nat. Phys. **11**, 690 (2015)
W.H. Matthaeus, Geophys. Res. Lett. **9**, 660 (1982)
K.T. Osman, K.H. Kiyani, W. H. Matthaeus, B. Hnat, S.C. Chapman, Yu.V. Khotyaintsev, Astrophys. J. Lett. **815**, L24 (2015)
T.D. Phan et al., Geophys. Res. Lett. **43**, 6060 (2016)
C. Pollock et al., Space Sci. Rev. **199**, 331 (2016)
L. Price, M. Swisdak, J.F. Drake, J. L. Burch, P.A. Cassak, R.E. Ergun, J. Geophys. Res. **122**, 11086 (2017)
L. Price, M. Swisdak, J.F. Drake, P. A. Cassak, J.T. Dahlin, R.E. Ergun, Geophys. Res. Lett. **43**, 6020 (2016)
P.L. Pritchett, J. Geophys. Res. **113**, A06210 (2008)
L. Rezeau, G. Belmont, R. Manuzzo, N. Aunai, J. Dargent, J. Geophys. Res. **123**, 227 (2018)
M.A. Shay et al., Geophys. Res. Lett. **43**, 4145 (2016)
M. Swisdak, J.F. Drake, L. Price, J.L. Burch, P.A. Cassak, T.D. Phan, Geophys. Res. Lett. **45**, 5260 (2018)
R.B. Torbert et al., Science **362**, 1391 (2018)
J.M. Webster et al., J. Geophys. Res. **123**, 4858 (2018)
M. Zhou et al., Astrophys. J. **870**, 34 (2019)

Chapter 8
ViDA: A Vlasov-DArwin Solver for Plasma Physics at Electron Scales

As discussed in previous chapters and especially in Chap. 5, numerical simulations have provided significant insights on the plasma dynamics at proton and sub-proton spatial scales, where proton kinetic effects are dominant, while electrons can be approximated as an isothermal fluid (hybrid framework) (Valentini et al. 2007). In this range of scales, both Particle-In-Cell (PIC) and Eulerian hybrid codes (see Sect. 5.2) have been extensively employed to investigate in detail a variety of physical phenomena. When it comes to investigating the electron scales it is found that only few numerical algorithms retaining both proton and electron kinetic physics are available at the moment. Most of them are PIC codes (Camporeale and Burgess 2017; González et al. 2019; Grošelj et al. 2017; Hesse et al. 2001; Horiuchi and Sato 1999; Markidis et al. 2010; Pritchett 2000; Ricci et al. 2002; Zeiler et al. 2002) which are able to capture also electron dynamics. However, PIC codes fail in providing a clean description of small-scale fluctuations, both in space, e.g. when computing the electric field behavior around the X-point, and in velocity space, for instance when calculating particle distribution functions, because of the intrinsic statistical noise (see Sect. 5.2). This problem can be overcome by using noise-free Eulerian algorithms, which, on the other hand, generally require a computational cost significantly larger than PIC codes (Büchner 2007, and references therein). The first attempts to describe plasma dynamics with Eulerian fully-kinetic codes has been reported in Sect. 5.2. This approach has recently become more affordable thanks to the improved supercomputer capabilities.

The ViDA (Vlasov-DArwin) code is a fully-kinetic Eulerian code that has been developed to advance in the understanding of electron scale processes in plasmas. In order to reduce the computational cost, the so-called Darwin approximation (Birdsall and Langdon 1985; Kaufman and Rostler 1971; Schmitz and Grauer 2006a) is applied. Within this approximation, all wave modes are retained except for light waves so that the constraint over the time step can be significantly relaxed. The ViDA

G. Cozzani, *Microphysics of Magnetic Reconnection in Near-Earth Space*, Springer Theses, https://doi.org/10.1007/978-3-030-56142-0_8

code integrates numerically the kinetic equations for a non-relativistic, globally-neutral plasma composed of ions and electrons. Equations are discretized on a fixed grid in phase space with periodic boundary conditions in the physical domain. ViDA originates from the hybrid Vlasov-Maxwell (HVM) code (Valentini et al. 2007) and it has been extended to include electron kinetic dynamics.

This chapter is meant to revisit the Darwin approximation, to discuss in detail the strategy of the numerical integration—namely the time splitting method for the distribution function of the two species and spectral method for the electromagnetic field—and to describe the algorithm design. Parallelization strategies and performances of the algorithm are also presented. As detailed in Pezzi et al. (2019), the code has been widely tested and it correctly reproduces the propagation of electrostatic Langmuir waves, whistler waves and Alfvén waves. Another test describes the onset of the electron Weibel instability, a plasma instability driven by the presence of a electron temperature anisotropy (Weibel 1959). Then, a test of the Vlasov-Darwin algorithm has been performed on a magnetic reconnection setup, including full ion and electron dynamics. Since the Thesis is devoted to the study of magnetic reconnection, this test in particular will be discussed in Sect. 8.4 while the reader is referred to (Pezzi et al. 2019) for details about the other tests. A large part of this chapter is presented in Pezzi et al. (2019). However, (i) Sect. 8.1 contains an extended discussion on the Darwin approximation, (ii) the splitting method is more largely discussed in Sect. 8.3.1 compared to the paper for sake of clarity in the context of the Thesis.

8.1 The Darwin Approximation

8.1.1 Darwin Approximated Lagrangian

According to the classical electromagnetic theory, the power P radiated by an electron is $P \propto (e^2/c^3)\mathrm{d}^2\mathbf{v}/\mathrm{d}t^2$. This dissipation cannot be represented in a Lagrangian form. It follows that if we want to approximate the Lagrangian of an electron which is moving in the potential due to another charged particle, we must not expect to be able to find a Lagrangian accurate beyond the terms of second order in the parameter v/c.

The fully relativistic Lagrangian $\mathcal{L}$ of a single particle moving in an electromagnetic field is

$$\mathcal{L} = -m_1 c^2 \beta_1 - e_1 \phi + \frac{e_1}{c}(\dot{\mathbf{r}}_1, \mathbf{A}) \tag{8.1}$$

where ϕ is the scalar potential and $\mathbf{A}$ is the vector potential

$$\mathbf{E} = -\nabla\phi - \frac{1}{c}\frac{\partial \mathbf{A}}{\partial t} \qquad \mathbf{B} = \nabla \times \mathbf{A}. \tag{8.2}$$

Here, e_1 is the particle charge, m_1 is the particle mass, $\mathbf{r}_1$ its position and $\beta_1 = \sqrt{1 + |\dot{\mathbf{r}}_1|/c^2}$. Now, suppose that there is a second particle of charge e_2 located at $\mathbf{r}_2$. The potentials ϕ and $\mathbf{A}$ can be written as follows

$$\phi = \left. \frac{e_2}{(r + (\dot{\mathbf{r}}_2, \mathbf{r}_2 - \mathbf{r}_1)/c)} \right|_{ret} \tag{8.3}$$

$$\mathbf{A} = \left. \frac{e_2}{c} \frac{\dot{\mathbf{r}}_2}{(r + (\dot{\mathbf{r}}_2, \mathbf{r}_2 - \mathbf{r}_1)/c)} \right|_{ret} \tag{8.4}$$

where $r = \sqrt{\mathbf{r}_1^2 + \mathbf{r}_2^2}$. These quantities must be evaluated at the retarded times. If the *delay* of the reception by e_1 of the signal departing from e_2 is τ then we have (Darwin 1920)

$$c^2\tau^2 = \mathbf{r}_2 - \mathbf{r}_1 - \dot{\mathbf{r}}_2\tau + \frac{1}{2}\ddot{\mathbf{r}}_2\tau^2 - \cdots \tag{8.5}$$

and we get

$$\tau = \frac{r}{c} - 2\tau\frac{(\dot{\mathbf{r}}_2, \mathbf{r}_2 - \mathbf{r}_1)}{c^2} + \frac{r}{2c^3}[\dot{\mathbf{r}}_2^2 + (\ddot{\mathbf{r}}_2, \mathbf{r}_2 - \mathbf{r}_1) + (\dot{\mathbf{r}}_2, \mathbf{r}_2 - \mathbf{r}_1)^2/r^2] - \cdots \tag{8.6}$$

Hence, the approximated expression for the potentials is

$$\phi = \frac{e_2}{r} + \frac{e_2}{2c^2}\left[\frac{\dot{\mathbf{r}}_2^2 + (\ddot{\mathbf{r}}_2, \mathbf{r}_2 - \mathbf{r}_1)}{r} + \frac{(\dot{\mathbf{r}}_2, \mathbf{r}_2 - \mathbf{r}_1)^2}{r^3}\right] \tag{8.7}$$

$$\mathbf{A} = \frac{e_2}{c}\frac{\dot{\mathbf{r}}_2}{r}. \tag{8.8}$$

In order to obtain the above expression for $\mathbf{A}$, we only used the first term of Eq. (8.6) since $\mathbf{A}$ is already multiplied by a factor $1/c$ in Eq. (8.1). The approximation of ϕ and $\mathbf{A}$ implies that the potentials are not evaluated at the retarded times i.e. the delays are neglected. Substituting Eqs. (8.7)–(8.8) in Eq. (8.1) we obtain

$$\mathcal{L} = -m_1c^2\beta_1 - \frac{e_1e_2}{r} - \frac{e_1e_2}{2c^2}\left[\frac{\dot{\mathbf{r}}_2^2 + (\ddot{\mathbf{r}}_2, \mathbf{r}_2 - \mathbf{r}_1) - 2(\dot{\mathbf{r}}_1, \dot{\mathbf{r}}_2)}{r} + \frac{(\dot{\mathbf{r}}_2, \mathbf{r}_2 - \mathbf{r}_1)^2}{r^3}\right] \tag{8.9}$$

Adding $-m_2c^2\beta_2$ and the total derivative $(e_1e_2/2c^2)[\mathrm{d}((\dot{\mathbf{r}}_2, \mathbf{r}_2 - \mathbf{r}_1)/r)/\mathrm{d}t]$ to Eq. (8.9) provides a Lagrangian which is symmetric to the exchange of e_1 and e_2. The two added terms have no effect on the equations of motion derived from Eq. (8.9). In addition, Eq. (8.9) describes also the motion of e_2 in the fields produced by e_1. Now, we rewrite Eq. (8.9) as follows

$$\mathcal{L} = -m_1 c^2 \beta_1 - m_2 c^2 \beta_2 - \frac{e_1 e_2}{r} + \frac{e_1 e_2}{2c^2}\left[\frac{(\dot{\mathbf{r}}_2, \dot{\mathbf{r}}_1)}{r} + \frac{(\dot{\mathbf{r}}_2, \mathbf{r}_2 - \mathbf{r}_1)(\dot{\mathbf{r}}_1, \mathbf{r}_2 - \mathbf{r}_1)}{r^3}\right] \tag{8.10}$$

and we expand also the term involving β_i consistently to the other term, i.e. up to v^2/c^2

$$-m_1 c^2 \beta_1 = -m_1 c^2 \frac{1}{\sqrt{1 + \frac{\dot{\mathbf{r}}_1^2}{c^2}}} = -m_1 c^2 + \frac{1}{2} m_1 \dot{\mathbf{r}}_1^2 + \frac{1}{8c^2} m_1 \dot{\mathbf{r}}_1^4 + \cdots \tag{8.11}$$

Substituting in Eq. (8.9) we get

$$\mathcal{L} = \frac{1}{2} m_1 \dot{\mathbf{r}}_1^2 + \frac{1}{8c^2} m_1 \dot{\mathbf{r}}_1^4 + \frac{1}{2} m_2 \dot{\mathbf{r}}_2^2 + \frac{1}{8c^2} m2 \dot{\mathbf{r}}_2^4 - \frac{e_1 e_2}{r} + \frac{e_1 e_2}{2c^2}\left[\frac{(\dot{\mathbf{r}}_2, \dot{\mathbf{r}}_1)}{r} + \frac{(\dot{\mathbf{r}}_2, \mathbf{r}_2 - \mathbf{r}_1)(\dot{\mathbf{r}}_1, \mathbf{r}_2 - \mathbf{r}_1)}{r^3}\right] \tag{8.12}$$

that can be easily generalized for N charged particles as

$$\mathcal{L} = \sum_{a=1}^{N} \frac{1}{2} m_a \dot{\mathbf{r}}_a^2 + \frac{1}{8c^2} m_a \dot{\mathbf{r}}_a^4 - \sum_{b<a} \frac{e_a e_b}{r_{ab}} + \sum_{b<a} \frac{e_a e_b}{2c^2}\left[\frac{(\dot{\mathbf{r}}_b, \dot{\mathbf{r}}_a)}{r_{ab}} + \frac{(\dot{\mathbf{r}}_b, \mathbf{r}_b - \mathbf{r}_a)(\dot{\mathbf{r}}_a, \mathbf{r}_b - \mathbf{r}_a)}{r_{ab}^3}\right]. \tag{8.13}$$

This is the approximated Lagrangian obtained by Darwin (1920). The scalar and vector potential read

$$\phi = \sum_{b<a} \frac{e_b}{r_{ab}} \tag{8.14}$$

$$\mathbf{A} = \sum_{b<a} \frac{e_b}{2c}\left[\frac{(\dot{\mathbf{r}}_b, \dot{\mathbf{r}}_a)}{r_{ab}} + \frac{(\dot{\mathbf{r}}_b, \mathbf{r}_b - \mathbf{r}_a)(\dot{\mathbf{r}}_a, \mathbf{r}_b - \mathbf{r}_a)}{r_{ab}^3}\right] \tag{8.15}$$

Following Krause et al. (2007), we write the potentials as functions of a continuum variable, the position $\mathbf{x}$:

$$\phi(\mathbf{x}, t) = \sum_b \frac{e_b}{|\mathbf{x} - \mathbf{r}_b|} \tag{8.16}$$

$$\mathbf{A}(\mathbf{x}, t) = \sum_b \frac{e_b}{2c} \frac{\dot{\mathbf{r}}_b + (\dot{\mathbf{r}}_b, \hat{\mathbf{r}}_b)\hat{\mathbf{r}}_b}{|\mathbf{x} - \mathbf{r}_b|} \tag{8.17}$$

8.1.2 Darwin Approximation in Plasma Physics

The Darwin approximation has found application in the contest of magnetised plasmas. As already mentioned in Sect. 5.2.3, when we consider the system of the Vlasov-Maxwell equations, the Courant-Friedrichs-Lewy (CFL) stability condition is introduced by the time integration of the fields. The CFL condition imposes a constraint on the time step Δt, since also the fast electromagnetic wave modes (the vacuum mode) has to be resolved on the grid. However, by assuming $v^2 \ll c^2$, we get a set of equations where the vacuum (transverse) modes are eliminated while all other wave modes are retained. Hence, a condition for the Darwin's approximation to be used is that the system is non-relativistic.

Let us consider the Maxwell equations (in cgs units):

$$\nabla \cdot \mathbf{E} = 4\pi e n \tag{8.18}$$

$$\nabla \cdot \mathbf{B} = 0 \tag{8.19}$$

$$\nabla \times \mathbf{E} = -\frac{1}{c}\frac{\partial \mathbf{B}}{\partial t} \tag{8.20}$$

$$\nabla \times \mathbf{B} = \frac{4\pi}{c}\mathbf{j} + \frac{1}{c}\frac{\partial \mathbf{E}}{\partial t} \tag{8.21}$$

According to the Helmholtz theorem (Griffiths 1962), the electric field can be decomposed into a longitudinal, irrotational component $\mathbf{E}_L$ and a transverse, solenoidal component $\mathbf{E}_T$. By definition, $\nabla \times \mathbf{E}_L = 0$ and $\nabla \cdot \mathbf{E}_T = 0$. Hence,

$$\nabla \cdot \mathbf{E}_L = 4\pi e n \tag{8.22}$$

$$\nabla \cdot \mathbf{B} = 0 \tag{8.23}$$

$$\nabla \times \mathbf{E}_T = -\frac{1}{c}\frac{\partial \mathbf{B}}{\partial t} \tag{8.24}$$

$$\nabla \times \mathbf{B} = \frac{4\pi}{c}\mathbf{j} + \frac{1}{c}\frac{\partial \mathbf{E}_L}{\partial t} + \frac{1}{c}\frac{\partial \mathbf{E}_T}{\partial t} \tag{8.25}$$

The Darwin approximation consists in neglecting the

$$\frac{1}{c}\frac{\partial \mathbf{E}_T}{\partial t}$$

term in Eq. (8.25) (Pezzi et al. 2019; Schmitz and Grauer 2006a) so that the final Maxwell-Darwin system of equations is

$$\nabla \cdot \mathbf{E}_L = 4\pi e n \tag{8.26}$$

$$\nabla \cdot \mathbf{B} = 0 \tag{8.27}$$

$$\nabla \times \mathbf{E}_T = -\frac{1}{c}\frac{\partial \mathbf{B}}{\partial t} \tag{8.28}$$

$$\nabla \times \mathbf{B} = \frac{4\pi}{c}\mathbf{j} + \frac{1}{c}\frac{\partial \mathbf{E}_L}{\partial t} \tag{8.29}$$

An advantage of using Darwin approximation is that Eqs. (8.28)–(8.29) can be solved without performing a time integration step and the fields are computed through the moments of the distribution at a given time. This point will be clarified in Sect. 8.2.

It is not trivial to see the equivalence between the Darwin approximation as described in Darwin (1920) and its formulation given in this section. This correspondence is clarified in the next section.

8.1.3 A Unified Approach

Following Krause et al. (2007), we want to show that the Darwin approximation as depicted by Darwin (1920) is equivalent to the formulation usually adopted in the field of plasma physics (see Sect. 8.1.2). The approximation, as formulated in Darwin (1920), basically consists in neglecting the delay in the evaluation of the retarded potentials. This is translated in plasma physics in neglecting the term involving $\mathbf{E}_T$ in the Ampère equation, Eq. (8.25). Krause et al. (2007) have shown that from neglecting the delay of the retarded potentials we can obtain the approximated Maxwell equations that are commonly called Maxwell-Darwin equations. We report here the quasistatic approach (Krause et al. (2007), Sect. II.A).

The Maxwell equations for the potentials are:

$$\nabla^2\phi + \frac{1}{c}\frac{\partial}{\partial t}(\nabla\cdot\mathbf{A}) = -4\pi e n \tag{8.30}$$

$$\nabla^2\mathbf{A} - \frac{1}{c^2}\frac{\partial^2\mathbf{A}}{\partial t^2} - \nabla\left(\nabla\cdot\mathbf{A} + \frac{1}{c}\frac{\partial\phi}{\partial t}\right) = -\frac{4\pi}{c}\mathbf{j} \tag{8.31}$$

We use the Coulomb gauge ($\nabla\cdot\mathbf{A} = 0$) and we obtain

$$\nabla^2\phi = -4\pi e n = -4\pi\rho \tag{8.32}$$

$$\nabla^2\mathbf{A} - \frac{1}{c^2}\frac{\partial^2\mathbf{A}}{\partial t^2} = \frac{1}{c}\frac{\partial\nabla\phi}{\partial t} - \frac{4\pi}{c}\mathbf{j} \tag{8.33}$$

The solutions for these equations in the case of unbounded region are known (Jackson 2002):

$$\phi_C(\mathbf{x}, t) = \int d^3x' \frac{\rho(\mathbf{x}', \mathbf{t})}{|\mathbf{x} - \mathbf{x}'|} \tag{8.34}$$

$$\mathbf{A}_C(\mathbf{x}, t) = \frac{1}{c}\int d^3x' \frac{(\mathbf{j}(\mathbf{x}', t') - \hat{\mathbf{r}}[\hat{\mathbf{r}}\cdot\mathbf{j}(\mathbf{x}', t')])_{ret}}{|\mathbf{x} - \mathbf{x}'|} +$$

$$+ c\int d^3x' \frac{1}{|\mathbf{x} - \mathbf{x}'|^3}\int_0^{r/c} d\tau\, \tau(3\hat{\mathbf{r}}[\hat{\mathbf{r}}\cdot\mathbf{j}(\mathbf{x}', t-\tau)] - \mathbf{j}(\mathbf{x}', t-\tau)) \tag{8.35}$$

where $r = |\mathbf{x} - \mathbf{x}'|$ and ret indicates that the quantities are evaluated at the retarded time $t' = t - r/c$. In the quasistatic approach the delay is neglected so we remove ret from the first part of the expression for $\mathbf{A}_C$ and we compute the second integral with the substitution $t - \tau \rightarrow t$. Therefore,

$$\mathbf{A}_C^{qs}(\mathbf{x},t) = \frac{1}{2c}\int d^3x' \frac{\mathbf{j}(\mathbf{x}',t) + \hat{\mathbf{r}}(\hat{\mathbf{r}}\cdot\mathbf{j}(\mathbf{x}',t))}{|\mathbf{x}-\mathbf{x}'|} \tag{8.36}$$

This is the continuum analog of the potential derived by Darwin (1920) in Eq. (8.17).

Now, the scalar potential ϕ_C satisfies the Poisson equation

$$\nabla^2\phi_C = -4\pi\rho. \tag{8.37}$$

Then, we want to compute the Ampère equation involving $\mathbf{A}_C^{qs}$. Since $\nabla\times\mathbf{B} = \nabla\times(\nabla\times\mathbf{A}) = -\nabla^2\mathbf{A} + \nabla(\nabla\cdot\mathbf{A})$ that in the Coulomb gauge reduces to $-\nabla^2\mathbf{A}$, Ampère equation becomes a Poisson equation for $\mathbf{A}$. We compute $-\nabla^2\mathbf{A}_C^{qs}$ from Eq. (8.36)

$$\begin{aligned}
\nabla^2\mathbf{A}_C^{qs} &= \frac{1}{2c}\int d^3x'\nabla^2_{\mathbf{x}}\left[\frac{\mathbf{j}(\mathbf{x}',t) + \hat{\mathbf{r}}(\hat{\mathbf{r}}\cdot\mathbf{j}(\mathbf{x}',t))}{|\mathbf{x}-\mathbf{x}'|}\right] = \\
&= \frac{1}{2c}\int d^3x' \,[\mathbf{j}(\mathbf{x}',t) + \hat{\mathbf{r}}(\hat{\mathbf{r}}\cdot\mathbf{j}(\mathbf{x}',t))]\nabla^2_{\mathbf{x}}\frac{1}{|\mathbf{x}-\mathbf{x}'|} + \\
&+ \frac{1}{2c}\int d^3x' \,\frac{1}{|\mathbf{x}-\mathbf{x}'|}\nabla^2_{\mathbf{x}}[\mathbf{j}(\mathbf{x}',t) + \hat{\mathbf{r}}(\hat{\mathbf{r}}\cdot\mathbf{j}(\mathbf{x}',t))] = \\
&= -\frac{4\pi}{c}\mathbf{j} + \frac{1}{c}\int d^3x'\frac{1}{r^3}[\mathbf{j}(\mathbf{x}',t) - 3\hat{\mathbf{r}}(\hat{\mathbf{r}}\cdot\mathbf{j}(\mathbf{x}',t))]
\end{aligned} \tag{8.38}$$

Now we decompose the vector $\mathbf{j}$ into its longitudinal and transverse component $\mathbf{j} = \mathbf{j}_L + \mathbf{j}_T$ (Griffiths 1962) that can be written in a general form as follows:

$$\mathbf{j}_T(\mathbf{x},t) = \frac{1}{4\pi}\nabla\times\nabla\times\int d^3x' \,\frac{\mathbf{j}(\mathbf{x}',t)}{r} \tag{8.39}$$

$$\mathbf{j}_L(\mathbf{x}',t) = -\frac{1}{4\pi}\nabla\int d^3x' \,\frac{\nabla_{\mathbf{x}'}\cdot\mathbf{j}(\mathbf{x}',t)}{r} \tag{8.40}$$

The longitudinal component may be expressed as

$$\mathbf{j}_L(\mathbf{x},t) = \frac{1}{4\pi}\int d^3x'\frac{1}{r^3}[\mathbf{j}(\mathbf{x}',t) - 3\hat{\mathbf{r}}(\hat{\mathbf{r}}\cdot\mathbf{j}(\mathbf{x}',t))] \tag{8.41}$$

If we substitute in Eq. (8.41) in the expression for $\nabla^2\mathbf{A}_C^{qs}$ we obtain:

$$\nabla^2\mathbf{A}_C^{qs} = -\frac{4\pi}{c}\mathbf{j} + \frac{4\pi}{c}\mathbf{j}_L = -\frac{4\pi}{c}\mathbf{j}_T \tag{8.42}$$

meaning that only the transverse currents participate to Ampère equation. Krause et al. (2007) have called this formulation the *quasistatic* approach. They have shown that this approach is equivalent to the so-called *operator approximation* approach where the structure of Eqs. (8.32)–(8.33) is modified from the beginning by neglecting the term $(1/c^2)\partial^2\mathbf{A}/\partial t^2$, which is the one responsible for the retard

$$\nabla^2\phi + \frac{1}{c}\frac{\partial}{\partial t}(\nabla\cdot\mathbf{A}) = -\,4\pi\rho \tag{8.43}$$

$$\nabla^2\mathbf{A} - \nabla(\nabla\cdot\mathbf{A}) - \frac{1}{c}\frac{\partial\nabla\phi}{\partial t} = -\frac{4\pi}{c}\mathbf{j} \tag{8.44}$$

With the ansatz $\nabla\cdot\mathbf{A} = 0$ and after calculations (see Krause et al. (2007), Sect. II.B) it is found that also in this case the vector potential satisfies the same equation as $\mathbf{A}_C^{qs}$, Eq. (8.42). Hence, the two methods are not equivalent *a priori* but they lead to the same result. Krause et al. (2007) have shown also that these approaches derive from an action principle and they re-obtain the action principle found in Darwin (1920). In addition, if Eqs. (8.37)–(8.42) are written in terms of the fields **E** and **B** using Eq. (8.32)–(8.33) we obtain the Maxwell equations written in the Darwin approximation as in Eqs. (8.26)–(8.29).

8.2 The Vlasov-Darwin Model of the ViDA Code

In this section we present the dimensionless equations that compose the Vlasov-Darwin model implemented in the ViDA code (Pezzi et al. 2019). We also show that the Maxwell-Darwin equations can be written as Helmholtz-like equations so that the only quantity that has to be evolved in time is the distribution function for each species. The system of equations, composed by the Vlasov equations self-consistently coupled to the Maxwell-Darwin equations, reads:

$$\partial_t f_\alpha + \mathbf{v}\cdot\nabla f_\alpha + \frac{Z_\alpha e}{m_\alpha}\left(\mathbf{E} + \frac{\mathbf{v}}{c}\times\mathbf{B}\right)\cdot\nabla_\mathbf{v} f_\alpha = 0 \tag{8.45}$$

$$\nabla\cdot\mathbf{E}_L = 4\pi e n \tag{8.46}$$

$$\nabla\cdot\mathbf{B} = 0 \tag{8.47}$$

$$\nabla\times\mathbf{E}_T = -\frac{1}{c}\partial_t\mathbf{B} \tag{8.48}$$

$$\nabla\times\mathbf{B} = \frac{1}{c}\partial_t\mathbf{E}_L + \frac{4\pi}{c}\mathbf{j} \tag{8.49}$$

where $f_\alpha(\mathbf{x}, \mathbf{v}, t)$ is the distribution function (DF) of the $\alpha = p, e$ species, and m_α and Z_α the mass and charge number of the α species, respectively. The plasma number density $n(\mathbf{x}, t)$ and the current density $\mathbf{j}(\mathbf{x}, t)$ are defined through the first two velocity moments of the particle DFs:

$$n = \sum_\alpha Z_\alpha n_\alpha = \sum_\alpha Z_\alpha \int d\mathbf{v}\, f_\alpha \tag{8.50}$$

$$\mathbf{j} = \sum_\alpha \mathbf{j}_\alpha = e \sum_\alpha Z_\alpha n_\alpha \mathbf{V}_\alpha = e \sum_\alpha Z_\alpha \int d\mathbf{v}\, \mathbf{v}\, f_\alpha \tag{8.51}$$

By normalizing these equations using a characteristic length $\bar{L}$, time $\bar{t}$, velocity $\bar{U} = \bar{L}/\bar{t}$, mass $\bar{m}$ and distribution function $f_{\alpha,0} = \bar{n}/\bar{U}^3$ (being $\bar{n} = \bar{L}^{-3}$ the equilibrium density), it is straightforward to get the dimensionless Vlasov-Darwin system of equations:

$$\partial_t f_\alpha + (\mathbf{v} \cdot \nabla)\, f_\alpha + \frac{Z_\alpha}{\mu_\alpha} (\mathbf{E} + \mathbf{v} \times \mathbf{B}) \cdot \nabla_\mathbf{v} f_\alpha = 0 \tag{8.52}$$

$$\nabla \cdot \mathbf{E}_L = \zeta^2 n \tag{8.53}$$

$$\nabla \cdot \mathbf{B} = 0 \tag{8.54}$$

$$\nabla \times \mathbf{E}_T = -\partial_t \mathbf{B} \tag{8.55}$$

$$\nabla \times \mathbf{B} = \bar{u}^2 \partial_t \mathbf{E}_L + \bar{u}^2 \zeta^2 \mathbf{j} \tag{8.56}$$

In Eqs. (8.52–8.56), the electric and magnetic fields are normalized to $\bar{E} = \bar{m}\bar{U}/e\bar{t}$ and $\bar{B} = \bar{m}c/e\bar{t}$, respectively. Note also that we set $k_B = 1$. Non-dimensional parameters are $\mu_\alpha = m_\alpha/\bar{m}$, $\bar{u} = \bar{U}/c$ and $\zeta = \bar{\omega}_p \bar{t}$, being $\bar{\omega}_p = \sqrt{4\pi e^2 \bar{n}/\bar{m}}$.

Equations (8.53–8.56) can be further simplified to obtain a set of Helmholtz-like equations that do not contain explicit time derivatives (see Schmitz and Grauer (2006a) for details). Hence, the system (8.52–8.56) can be re-written as follows

$$\partial_t f_\alpha + (\mathbf{v} \cdot \nabla)\, f_\alpha + \frac{Z_\alpha}{\mu_\alpha} (\mathbf{E} + \mathbf{v} \times \mathbf{B}) \cdot \nabla_\mathbf{v} f_\alpha = 0 \tag{8.57}$$

$$\nabla^2 \phi = -\zeta^2 \sum Z_\alpha n_\alpha \qquad \mathbf{E}_L = -\nabla \phi \tag{8.58}$$

$$\nabla^2 \mathbf{B} = -\bar{u}^2 \zeta^2 \nabla \times \mathbf{j} \tag{8.59}$$

$$\nabla^2 \hat{\mathbf{E}}_T - \bar{u}^2 \zeta^2 \sum_\alpha \frac{Z_\alpha^2 n_{\alpha,0}}{\mu_\alpha} \hat{\mathbf{E}}_T = \bar{u}^2 \zeta^2 \left[-\nabla \cdot \sum_\alpha Z_\alpha \langle \mathbf{v}\mathbf{v} \rangle_\alpha + \right.$$
$$\left. + \sum_\alpha \frac{Z_\alpha^2}{\mu_\alpha} (n_\alpha \mathbf{E}_L + \langle \mathbf{v} \rangle_\alpha \times \mathbf{B}) \right] \tag{8.60}$$

$$\nabla^2 \Theta = \nabla \cdot \hat{\mathbf{E}}_T \qquad \mathbf{E}_T = \hat{\mathbf{E}}_T - \nabla \Theta \tag{8.61}$$

$$\nabla \cdot \mathbf{B} = 0 \tag{8.62}$$

where $\langle h \rangle_\alpha = \int d\mathbf{v} f_\alpha h$. Note that in Eq. (8.60) we have omitted a term $\bar{u}^2 \nabla \partial_{tt} \phi$ which could generate, in principle, an irrotational component, and we have introduced Eqs. (8.61) to preserve the solenoidality of $\mathbf{E}_T$ (Schmitz and Grauer 2006a). The spatial dependence of n_α on the left-hand side of Eq. (8.60) has been neglected ($n_\alpha \simeq n_{\alpha,0}$) to let coefficients be constant (Valentini et al. 2007).

It is worth noticing that, since Darwin equations are a set of Helmholtz-like equations, initial perturbations have to be introduced through the particle DFs. This represents a difference with respect to standard codes where also magnetic perturbations can be introduced and it is important for the design of the initial condition, as it will be addressed in Sect. 9.1.

At variance with previous models (Valentini et al. 2007), the Darwin model retains the longitudinal component of the displacement current. In this viewpoint, the Darwin system is closer to the full Maxwell system with respect to models where the displacement current is completely neglected.

8.3 ViDA Algorithm and Code Design

The Vlasov equation for each species is integrated numerically by employing the time splitting method first proposed by Cheng and Knorr (1976) to solve the Vlasov-Poisson system in the electrostatic limit and later extended to the full electromagnetic case (Mangeney et al. 2002). Maxwell-Darwin equations are solved through standard Fast Fourier Transform (FFT) algorithms.

8.3.1 *The Splitting Scheme*

As already mentioned, the ViDA code has been developed as the extension of the hybrid Vlasov-Maxwell code presented in Valentini et al. (2007). In particular, the splitting algorithm used in the ViDA code for the distribution functions of the two species is exactly the one implemented in the hybrid Vlasov-Maxwell code for the ion distribution function. In this section we revise the splitting scheme addressed in Valentini et al. (2007), Mangeney et al. (2002).

The Vlasov equation (8.57) is a multidimensional advection equation. The characteristics of Eq. (8.57) correspond to the particle trajectories (we suppose $\mu_\alpha = 1$ and $Z_\alpha = -1$ for simplicity)

$$\frac{d\mathbf{x}}{dt} = \mathbf{v}, \qquad \frac{d\mathbf{v}}{dt} = -(\mathbf{E} + \mathbf{v} \times \mathbf{B}) \tag{8.63}$$

and they describe a Hamiltonian flow $\mathcal{T}^t$ in phase space so that a point in phase space $z = (\mathbf{x}, \mathbf{v})$ evolves as follows

$$z(z_0, t) = \mathcal{T}^t z_0.$$

The Hamiltonian flow $\mathcal{T}^t$ is reversible and preserves the volume in phase space.

Now, the evolution of the distribution function $f(z, t)$ can be written in terms of $\mathcal{T}^t$

$$f(z, t) = f_0(\mathcal{T}^{-t} z) = T^t f_0(z)$$

where f_0 is the distribution function at $t = 0$ and T is the operator that evolves the distribution function. In order to provide an approximation for T^t the splitting method is used.

We consider the electrostatic case to introduce the splitting method. In that case, the particle motion is described by the Hamiltonian

$$H = \frac{v^2}{2} + \phi(x)$$

and the Vlasov equation can be written in terms of H

$$\frac{\partial f}{\partial t} = -[H, f] \equiv \Lambda f$$

so that the evolution operator is $T^t = \Lambda t$. In this case, the Hamiltonian can be split into two parts ($H_1 = v^2/2$ and $H_2 = \phi(x)$) so that

$$\frac{\partial f}{\partial t} = -[H, f] = -[H_1 + H_2, f] = -[H_1, f] - [H_2, f] \equiv \Lambda_1 t + \Lambda_2 t$$

The two operators $e^{\Lambda_1 t}$ and $e^{\Lambda_2 t}$ correspond to translations in regular space or velocity space so that

$$e^{\Lambda_1 t} f(x, v) = f(x - vt, v), \qquad e^{\Lambda_2 t} f(x, v) = f(x, v + (\partial\phi/\partial x)t) = f(x, v - E(x)t).$$

It can be shown that

$$e^{\Lambda t} = \lim_{N\to\infty} \left[\exp\left(\frac{\Lambda_2 t}{2N}\right) \exp\left(\frac{\Lambda_1 t}{N}\right) \exp\left(\frac{\Lambda_2 t}{2N}\right)\right]^N. \tag{8.64}$$

For a finite time step $\tau = 1/N$ the second order approximation of $e^{\Lambda t}$ is

$$e^{\Lambda\tau} = \exp\left(\frac{\Lambda_2 \tau}{2}\right) \exp(\Lambda_1 \tau) \exp\left(\frac{\Lambda_2 \tau}{2}\right) + O(\tau^3). \tag{8.65}$$

The operator $e^{\Lambda\tau}$ corresponds to an approximation of the Hamiltonian flow T^t to the second order

$$x_\tau = T^t x_0 = x + \tau v + \frac{\tau^2}{2} E\left(x + v\frac{\tau}{2}\right), \qquad v_\tau = T^t v_0 = v + \tau E\left(x + v\frac{\tau}{2}\right). \tag{8.66}$$

This transformation is symplectic, i.e. it conserves the volume in the phase space ($dxdv = dx_\tau dv_\tau$). If Λ_1 and Λ_2 commute we can further write

$$e^{\Lambda\tau} = \exp[(\Lambda_1 + \Lambda_2)\tau].$$

However, Eq. (8.65) is valid for every decomposition of the Hamiltonian.

The following step consists in generalizing this approach to a case with both electric and magnetic field arbitrarily directed. Analogously to Eq. (8.65), the time advance operator is actually formed by seven translation operators for the velocity space. Following Mangeney et al. (2002), we define

$$\Theta_x = (E_x + (\mathbf{v} \times \mathbf{B})_x)\frac{\partial}{\partial v_x}, \quad \Theta_y = (E_y + (\mathbf{v} \times \mathbf{B})_y)\frac{\partial}{\partial v_y}, \quad \Theta_z = (E_z + (\mathbf{v} \times \mathbf{B})_z)\frac{\partial}{\partial v_z}$$

and

$$\Omega_z(\tau) = \exp\left(\frac{\Theta_x \tau}{2}\right)\exp(\Theta_y \tau)\exp\left(\frac{\Theta_x \tau}{2}\right) \tag{8.67}$$

and

$$\Lambda_{xyz} = -\left(v_x\frac{\partial}{\partial x} + v_y\frac{\partial}{\partial y} + v_z\frac{\partial}{\partial z}\right) \tag{8.68}$$

which is the operator of translation in the regular space.

Hence, the time advance of the distribution function is performed as follow

$$\begin{aligned} f(\mathbf{x}, \mathbf{v}, \tau) &= \exp\left(\frac{\Lambda_{xyz}\tau}{2}\right)\left[\Omega_z\left(\frac{\tau}{2}\right)\exp(\Theta_z\tau)\Omega_z\left(\frac{\tau}{2}\right)\right]\exp\left(\frac{\Lambda_{xyz}\tau}{2}\right) f(\mathbf{x}, \mathbf{v}, 0) = \\ &= \exp\left(\frac{\Lambda_{\mathbf{x}}\tau}{2}\right)\exp(\Lambda_{\mathbf{v}}\tau)\exp\left(\frac{\Lambda_{\mathbf{x}}\tau}{2}\right) f(\mathbf{x}, \mathbf{v}, 0) \end{aligned} \tag{8.69}$$

where $\Lambda_{\mathbf{x}}$ is the advection operator in regular space and $\Lambda_{\mathbf{v}}$ is the operator in velocity space. To maintain second-order accuracy in τ, the electric and magnetic fields which are used to calculate the force in Θ_i $(i = x, y, z)$ are determined self consistently by solving the Maxwell-Darwin equations using currents and charges calculated with the translated distribution function $\exp\left(\frac{\Lambda_{\mathbf{x}}\tau}{2}\right) f$ (after the first half-time step).

Note that Eq. (8.67) has been defined as depending upon Θ_i and Θ_j with $i = x$, $j = y$ but this choice it arbitrary for any $i, j = x, y, z$.

Within the ViDA code, both spatial and velocity advections have been performed through a third-order Van Leer scheme. The details of this scheme are not detailed here but they are fully addressed in Mangeney et al. (2002, Sect. 3.3.1 and Appendix III).

8.3.2 *Structure of the Algorithm of ViDA*

The structure of ViDA for advancing the distribution functions from the time instant t_n to the time instant t_{n+1} is the following (see Mangeney et al. (2002) for the definition of the advection operators in physical and velocity space, respectively $\Lambda_{\mathbf{x}}$ and $\Lambda_{\mathbf{v}}$):

1. Performing the physical-space advection: $\tilde{f}_\alpha^n = \Lambda_{\mathbf{x}} f_\alpha^n$, where f_α^n is the α-species distribution function at the time instant t_n;
2. Computing the moments of $\tilde{f}_\alpha^n$: n_α, $\langle \mathbf{v} \rangle_\alpha$ and $\langle \mathbf{vv} \rangle_\alpha$ and evaluating the electromagnetic fields $\mathbf{E}_L$, $\mathbf{E}_T$ and $\mathbf{B}$, solving Eqs. (8.58–8.62) through standard Fast Fourier Transform (FFT) algorithms;
3. Performing the velocity-space advection $f^{n+1} = \Lambda_{\mathbf{v}} \tilde{f}_\alpha^n$.

A check on the solenoidality of $\mathbf{B}$ and $\mathbf{E}_T$ is implemented at each time step.

The phase space domain is discretized as follows. The physical space $D_{\mathbf{x}} = [0, L_x] \times [0, L_y] \times [0, L_z]$ is discretized with $N_x \times N_y \times N_z$ gridpoints. Boundary conditions are periodic in all directions. The velocity space $D_{\mathbf{v},\alpha} = [-v_{\alpha,x}^{max}, v_{\alpha,x}^{max}] \times [v_{\alpha,y}^{max}, v_{\alpha,y}^{max}] \times [-v_{\alpha,z}^{max}, v_{\alpha,z}^{max}]$ is discretized by $(2N_{\alpha,v_x} + 1) \times (2N_{\alpha,v_y} + 1) \times (2N_{\alpha,v_z} + 1)$ grid points. Different grids in phase space for the distribution function of protons and electrons are used. This is justified since the Vlasov equations for protons and electrons are only coupled by the electromagnetic field. Velocity-space boundary conditions impose $f_\alpha(|v_i| > v_{\alpha,i}^{max}) = 0$ $(i = x, y, z)$. In order to ensure mass conservation, $v_{\alpha,i}^{max}$ is typically set to be a large multiple of the thermal speed $v_{th,\alpha} = \sqrt{T_\alpha / m_\alpha}$.

The ViDA algorithm has been designed in such a way that the user can select (i) different normalizations of the model equations, (ii) the possibility of setting motionless protons and (iii) different dimensionalities of the physical-space domain ($1D$, $2D$, or $3D$), the velocity-space domain being always three-dimensional ($3V$). Spatial vectors always have three components but they can depend upon one, two or three spatial variables, depending on the physical-space dimensionality.

8.3.3 Normalizations of the Vlasov-Darwin Equations

In order to normalize Eqs. (8.57–8.62), three possible choices have been implemented in ViDA:

1. *Electrostatic* normalization. Characteristic quantities are: length $\bar{L} = \lambda_{D,e}$, time $\bar{t} = \omega_{p,e}^{-1}$, velocity $\bar{U} = v_{th,e}$ and mass $\bar{m} = m_e$. Here $\lambda_{D,e} = \sqrt{T_e / 4\pi n_e e^2}$, $\omega_{p,e} = \sqrt{4\pi n_e e^2 / m_e}$, $v_{th,e} = \sqrt{T_e / m_e} = \lambda_{D,e}\omega_{p,e}$ and m_e are the electron Debye length, the electron plasma frequency, the electron thermal speed and the electron mass, respectively. This normalization is appropriate for describing phenomena occurring at electron scales, such as the propagation of electrostatic plasma waves.
2. *Electromagnetic* normalization. Characteristic quantities are: length $\bar{L} = d_e$, time $\bar{t} = \omega_{p,e}^{-1}$, velocity $\bar{U} = c$ and mass $\bar{m} = m_e$, where $d_e = c/\omega_{p,e}$ is the electron skin depth. This normalization can be adopted for describing electromagnetic phenomena, where both protons and electrons are involved, such as magnetic reconnection and plasma turbulence at kinetic scales.
3. *Hybrid* normalization. Characteristic quantities are: length $\bar{L} = d_p$, time $\bar{t} = \Omega_{c,p}^{-1}$, velocity $\bar{U} = v_A$ and mass $\bar{m} = m_p$. In previous expressions $\Omega_{cp} = eB_0/$

$m_p c$, $v_A = B_0/\sqrt{4\pi n_p m_p}$, $d_p = v_A/\Omega_{cp}$ and m_p are the proton cyclotron frequency, the proton Alfvén speed, the proton skin depth and the proton mass, respectively. This normalization is useful for investigating the turbulent cascade in the sub-proton range, where electron physics starts to play a role.

8.3.4 Parallelization and Computational Cost

Since the computational effort needed to evolve the Vlasov-Darwin system of equations is significant, the ViDA code is massively parallelized. The standard parallelization protocols are the Message Passing Interface (MPI) and the Open Multi-Processing (openMP) interface that supports shared memory multiprocessing programming. The parallelization of the ViDA code is based on the synergy between the MPI and OpenMP paradigms. The MPI paradigm is adopted to parallelize the computational domain of the regular space for particle distribution functions, their moments and the electromagnetic field so that each MPI thread accesses a portion of phase space (cubic cells in $3D$ and squared cells in $2D$), composed by a sub-portion of the regular space and by the whole velocity space. Within each MPI thread, the OpenMP directives are adopted to parallelize the velocity-space cycles.

Preliminary performance tests of ViDA have been implemented on the Marconi-KNL cluster at the CINECA supercomputing center (Casalecchio di Reno (BO), Italy) and they are presented in Pezzi et al. (2019). The Marconi cluster is equipped with 3600 Lenovo Adam Pass nodes, interconnected through the Intel OmniPath network and each one composed by 1KNL processor (68 cores, 1.40GHz), formally 96 GB of RAM (effective 83 GB) and 16 GB of MCDRAM. The tests have been performed on a simple equilibrium configuration (Maxwellian DFs with no perturbations). It is found that, for 2D simulations and within the current parallelization, the best performance is achieved with 32 MPI threads and 2 OpenMP tasks per node on a KNL system. These preliminary tests show a reasonable parallel efficiency on KNL architecture, at least up to some hundreds cores.

Concerning the computational costs, the ViDA code is about twice as computationally expensive as the HVM code (Valentini et al. 2007), which has been recently used for 3D simulations of plasma turbulence (see for instance Cerri et al. (2018)). More specifically, the reconnection run presented in Sect. 8.4—which is the most expensive test performed in Pezzi et al. (2019) in terms of required computational resources—has a cost of slightly less than 0.1 Mh on Marconi supercomputer using 16 nodes and 512 MPI processes. On the other hand, being ViDA a code for a new piece of physics, it is difficult to foresee for the exact cost of a 3D reconnection (or turbulence) run because the numerical and physical parameters, as well as the duration of the run, can vary significantly with respect to the standard ones used with the HVM code. Based on the experience with the HVM code, it can be suggested that a high resolution 3D run of magnetic reconnection focusing on the electron physics would take from a few to a few tens of Mh.

8.4 Dynamics of Magnetic Reconnection

This section presents results of a magnetic reconnection simulation. Vlasov simulations of magnetic reconnection represent a strong numerical challenge because of the huge memory and CPU time required by Eulerian algorithms. This approach, if successful, would certainly provide a crucial contribution to the understanding of the magnetic reconnection process especially at electron scales, thanks to the fact that Eulerian algorithms allow for an almost noise-free description of fields and particle distribution functions. A noise-free description is crucial to properly understand which electromagnetic fluctuations contribute to the reconnection electric field in the form of anomalous resistivity and how distribution functions are modified leading to electron heating. Indeed, the contribution of the terms of Ohm's law to the reconnection electric field is investigated in Chap. 9 by means of simulations performed with the ViDA code.

I have performed a 2D–3V (two dimensional in physical space and three dimensional in velocity space) symmetric magnetic reconnection simulation. The initial condition of our simulation is the one adopted in the Geospace Environmental Modeling (GEM) magnetic reconnection challenge (Birn et al. 2001), in order to allow for a direct comparison to previous studies (Birn et al. 2001; Schmitz and Grauer 2006b). For this reason, I have also chosen the *hybrid* normalization so that the characteristic length $\bar{L}$, time $\bar{t}$, velocity $\bar{U}$ and mass $\bar{m}$ are $\bar{L} = d_p$, $\bar{t} = \Omega_{c,p}^{-1}$, $\bar{U} = v_A$ and $\bar{m} = m_p$ respectively.

The equilibrium is set by adapting the Harris equilibrium (Harris 1962) to the periodic boundary conditions in the spatial domain. In particular, the component of the magnetic field $B_x(y)$ corresponding to the double current sheet profile reads:

$$B_x(y) = B_0\left[\tanh\left(\frac{y - L_y/2}{L_1}\right) - \tanh\left(\frac{y}{L_2}\right) - \tanh\left(\frac{y - L_y}{L_2}\right)\right]. \tag{8.70}$$

This profile is characterized by the presence of two gradients (corresponding to the current sheets) varying as an hyperbolic tangent and located at $y = L_y/2$ and $y = 0$ (and so at $y = L_y$) where L_y is the length of the spatial domain in the y direction. The first hyperbolic tangent is the one defined in Harris (1962) and L_1 is the corresponding current sheet thickness. The second and third hyperbolic tangent in Eq. (8.70) have been included to satisfy the spatial periodicity; the value of L_2 is taken sufficiently large compared to L_1 to slow down the development of reconnection in the second current sheet with respect to the main one. In this way, reconnection initially develops only at the steeper current sheet while the less steep one stays basically inactive during the time interval that is taken into account. Figure 8.1 shows the evolution of the z component of the magnetic field in the whole domain of the simulation for four different times. When B_z exhibits an enhanced quadrupolar pattern we can infer that reconnection is ongoing. At the initial time, the GEM-like perturbation is imposed on both the current sheet. During the time interval in which reconnection is ongoing in the current sheet in the upper part of the reconnection

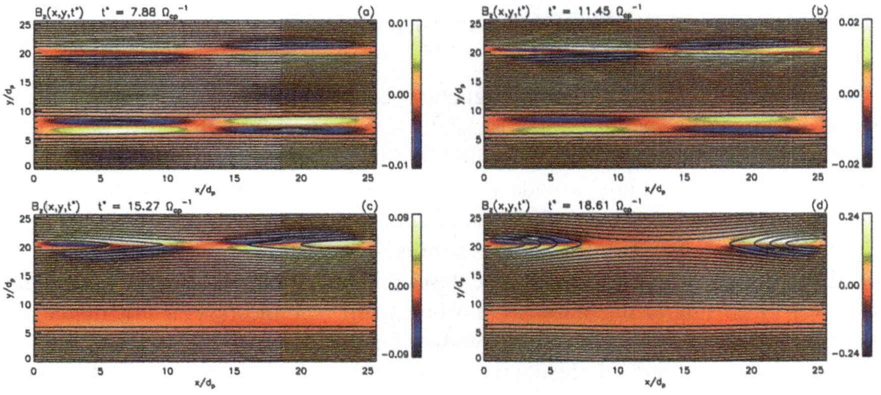

Fig. 8.1 Contour plots of B_z at four different times **a** $t^* = 7.88\ \Omega_{c,p}^{-1}$; **b** $t^* = 11.45\ \Omega_{c,p}^{-1}$; **c** $t^* = 15.27\ \Omega_{c,p}^{-1}$; **d** $t^* = 18.61\ \Omega_{c,p}^{-1}$. The domain has been shifted of 7.5 d_p in the y direction in order to better visualize both current sheets. The contour lines of the magnetic flux ψ are superposed

box (panels (c) and (d)), the lower current sheet is still not active and B_z there is negligible compared to the Hall magnetic field of the steeper current sheet.

The electron and ion temperature are set uniform at the initial time and the density $n(y)$ is defined in order to satisfy pressure balance. Then, from Eq. (8.49) and considering $\partial_t \mathbf{E}_L = 0$ at the initial time, we get the initial current density $\mathbf{j} = (0,\ 0,\ j_z(y))$.

Following the prescriptions of the Harris equilibrium we get, in normalized units,

$$n_0(T_e + T_p) = \frac{B_0^2}{2} \tag{8.71}$$

$$\frac{u_{e,0}}{T_e} = -\frac{u_{p,0}}{T_p} \tag{8.72}$$

$$\frac{j_z(y)}{n(y)} \equiv u_0 = u_{p,0} - u_{e,0} \tag{8.73}$$

Equation (8.72) corresponds to the no charge separation condition of the Harris equilibrium so that quasi-neutrality is imposed, $n_e(y) = n_p(y) = n(y)$. In other words, the electric field is zero at the initial time. Moreover, from Eqs. (8.72)–(8.73) we have:

$$u_{e,0} = -\frac{u_0}{1 + T_p/T_e} \tag{8.74}$$

$$u_{p,0} = \frac{u_0}{1 + T_e/T_p} \tag{8.75}$$

More details about the design of the initial condition for the magnetic reconnection run performed with the ViDA code are discussed in Sect. 9.1.

It is worth to point out that this is not an exact Vlasov kinetic equilibrium. In particular, it differs from the equilibrium presented by Harris since in this simulation the spatial domain is periodic in the varying y-direction. On the other hand, the initial configuration is in force balance and I have checked that the initial equilibrium is not significantly affected by, for example, ballistic effects within the time scale of reconnection considered here.

As for the GEM challenge (Birn et al. 2001), fluctuations are superposed to the initial magnetic field in order to obtain a single magnetic island at the center of the space domain at the initial time. In particular, $\delta\mathbf{B} = \nabla\delta\psi \times \hat{z}$ and

$$\delta\psi(x, y) = \psi_0 \cos(2\pi x/L_x)\cos(2\pi y/L_y) \tag{8.76}$$

where, L_x and L_y are the lengths of the spatial domain in x and y direction respectively. According to GEM challenge, in scaled units, ψ_0 is set to 0.1.

By using the relation $\delta\mathbf{B}(x, y) = \nabla\delta\psi(x, y) \times \hat{z}$ and Eq. (8.59), we derive the expression for the current density fluctuations $\delta\mathbf{j}(x, y)$ consistent with $\delta\psi(x, y)$. In particular, it is possible to define $\delta\mathbf{j}(x, y) = (0, 0, \delta j_z(x, y))$. Finally, the initial electron and proton distribution functions are shifted Maxwellian distributions with drift velocities along the z direction and temperature T_e and T_p. A uniform background density n_∞ is added for both electrons and protons. The background plasma does not affect the pressure balance but it guarantees that the distribution functions are well characterized in the whole simulation box.

The phase space has been discretized with $N_x \times N_y = 512 \times 512$ gridpoints in the spatial domain, $N_{e,v_x} \times N_{e,v_y} \times N_{e,v_z} = 41 \times 41 \times 81$ gridpoints in the velocity domain for electrons and $N_{p,v_x} \times N_{p,v_y} \times N_{p,v_z} = 31 \times 31 \times 31$ gridpoints in the velocity domain for protons. Also $v_e^{max} = 5\ v_{th,e}$ and $v_p^{max} = 5\ v_{th,p}$, where the normalized $v_{th,p}$ is set to 1. Other simulation parameters are $L_1 = 0.5d_p$, $L_2 = 2.5d_p$, $m_p/m_e = 25$, $n_\infty = 0.2$, $T_e/T_p = 0.2$, $L_x = L_y = 25.6d_p$, $\Delta y = \Delta x = L_x/N_x = 0.05\,d_p = 0.25\,d_e$. Also, I set $B_0 = 1$ and $n_0 = 1$ where B_0 is the asymptotic magnetic field while n_0 is the value of the density at the peak in the center of the current sheet. All parameters are chosen to be as close as possible to the simulation parameters listed in Birn et al. (2001).

Figure 8.2 shows the evolution of the reconnected magnetic flux given by the difference $\Delta\psi$ between the magnetic flux ψ evaluated at the X-point and at the O-point. Accordingly to the initial perturbation, the X and the O point are initially located at $(L_x/2, L_y/2)$ and $(0, L_y/2)$ and their location does not significantly change throughout the simulation run. The behavior of $\Delta\psi$ is very similar to the evolution of the reconnected magnetic flux in Ref. (Birn et al. 2001). Reconnection evolves with a reconnected magnetic flux that remains close to zero until $t \sim 15\,\Omega_{c,p}^{-1}$, when a sharp increase is observed. Then, the reconnection rate stays relatively constant until the reconnected flux begins to saturate at $t \sim 30\,\Omega_{c,p}^{-1}$.

Figure 8.3 shows the contour plots of the out of plane magnetic field B_z (a), of the electron current density in the z-direction $j_{e,z}$ (b), of the proton current density in the z-direction $j_{p,z}$ (c) and of the electron number density n_e (d). In each panel, the contour lines of the magnetic flux ψ are superposed. B_z exhibits the typical Hall

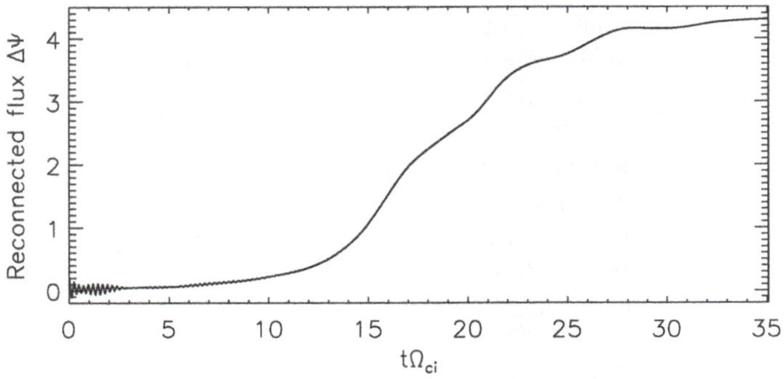

Fig. 8.2 Time evolution of the reconnected magnetic flux $\Delta\psi$. Reproduced with permission. © Cambridge University Press 2019

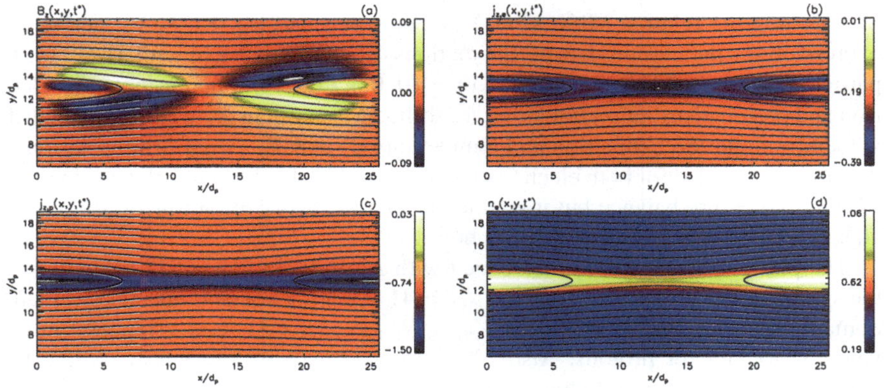

Fig. 8.3 Contour plots of B_z (**a**); out-of-plane electron current density $j_{e,z}$ (**b**); out-of-plane proton current density $j_{p,z}$ (**c**); and electron number density n_e (**d**). The quantities are shown at the time $t^* = 15.27\,\Omega_{c,p}^{-1}$. At that time $\Delta\psi = 1.18$. All the panels are zoomed in y in the interval $[6\,d_p, 19\,d_p]$. The contour lines of the magnetic flux ψ are superposed. Reproduced with permission. © Cambridge University Press 2019

quadrupolar pattern usually observed during symmetric magnetic reconnection. This magnetic signature indicates that the protons are demagnetized while the electrons are still frozen to the magnetic field. The difference in their dynamics produces the out-of-plane B_z (Mandt et al. 1994; Uzdensky and Kulsrud 2006) as detailed in Sect. 2.4. The quadrupolar structure that we find is analogous to the one obtained with other kinetic codes, both Eulerian (Schmitz and Grauer 2006b, see Fig. 2) and Lagrangian (Pritchett 2001, see Plate 1(b)). Note that the $j_{p,z}$ pattern closely follows the density pattern ($n_e \simeq n_p$) so that $j_{p,z}$ is depleted at the X-point while it reaches its maximum value within the magnetic island. On the other hand, $j_{e,z}$ is enhanced at the X-point and the region of strong $j_{e,z}$ is elongated along x. Away from the X-point, $j_{e,z}$ splits into two branches that identify the separatrices, as it was also

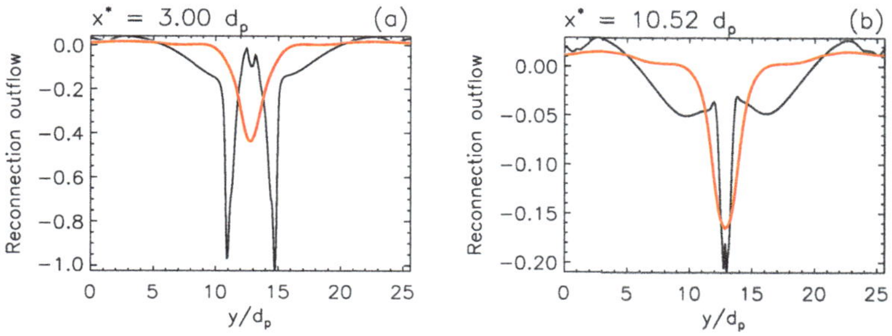

Fig. 8.4 **a** x component of the electron velocity u_e (black line) and of the proton velocity u_p (red line) at $x^* = 3.00\,d_p$ and **b** at $x^* = 10.52\,d_p$. The quantities are shown at the time $t^* = 18.13\,\Omega_{c,p}^{-1}$. Reproduced with permission. © Cambridge University Press 2019

observed by Shay et al. (2001) (see Fig. 6(d)). The electron current at the X line has a thickness comparable to d_p which corresponds to 5 d_e. The maximum value of the normalized B_z is 0.09 while the maximum values of $j_{p,z}$ and $j_{e,z}$ are 0.39 and 1.49, respectively. These values are overall slightly smaller than the values found in a similar Vlasov-Darwin simulation described in Ref. (Schmitz and Grauer 2006b).

Figure 8.4 shows the reconnection outflow of protons and electrons at $t^* = 18.13\ \Omega_{c,p}$. In particular, we note that at $x = 3\ d_p$ (panel (a)), corresponding to a distance of 9.8 d_p from the X-point located at $L_x/2 = 12.8\ d_p$, the electron velocity is characterized by two peaks corresponding to the separatrices, while the proton velocity is concentrated in the center of the outflow region and it reaches lower values, as expected. The presence of the two peaks is consistent with the $j_{e,z}$ pattern shown in Fig. 8.3b. Figure 8.4b shows the same quantities of Fig. 8.4a at a distance of 2.3 d_p from the X-point where the outflow is still developing and we note that $u_{e,x}$ is rather similar in shape and value to $u_{p,x}$.

8.5 Conclusions

This chapter presents a fully-kinetic code (ViDA) based on a Vlasov-Darwin algorithm, where only light waves are excluded in order to relax the constraint on the timestep advancement. This approach is particularly suited for the investigation of the kinetic dynamics from sub-ion scales down to the electron kinetic scales d_e and to the Debye length λ_D. As typically the case for space plasmas, but often also in the laboratory, inter-particle collisions are not described, since collisional scales are assumed to be smaller than other characteristics dynamical scales.

ViDA has been tested against several waves modes, in particular Alfvén, whistlers and plasma waves. The development of the Weibel instability and reconnection, both in a regime where the main dynamics is driven by the electrons, has been also reproduced. These tests represent typical regimes of interest for studying the electron scale kinetic dynamics representing at today a strong computational challenge and a frontier problem for the understanding of the electron plasma physics. These tests have been extensively detailed in Pezzi et al. (2019).

Being this Thesis focused on the reconnection process, I reported results of a 2D symmetric magnetic reconnection simulation in Sect. 8.4. The simulation parameters have been chosen to be close to the parameters used in the GEM challenge (Birn et al. 2001) in order to easily compare the obtained results with previous studies. The typical signatures of ongoing reconnection are reproduced, in particular the quadrupolar Hall magnetic field. This magnetic reconnection run has been designed to test the new code, another run with higher resolution and with a normalization more suited to study electron scales (the *electromagnetic* normalization of Sect. 8.3.3) is discussed in the next chapter.

One of the main future objectives of ViDA will be the study of the structure and dynamics of the electron diffusion region, including the role of anomalous resistivity in the Ohm's law and the mechanisms of electron heating, which are among the main targets of satellite MMS data analysis (Genestreti et al. 2018; Torbert et al. 2016a). Last but not least, the ViDA code will be used for the study of the plasma turbulent dynamics focusing on the problem of the "dissipative" scale, of primary interest in the context of the solar wind turbulent heating at kinetic scales (Vaivads et al. 2016).

References

C.K. Birdsall A.B. Langdon, *Plasma Physics Via Computer* (McGraw-Hill, Inc., New York, 1985)
J. Birn et al., J. Geophys. Res. **106**, 3715 (2001)
J. Büchner, in *Advanced Methods for Space Simulations*, ed. by H. Usui, Y. Omura (TERRAPUB, Tokyo, 2007), p. 23
E. Camporeale, D. Burgess, J. Plasma Phys. **83**, 535830201 (2017)
S.S. Cerri, M.W. Kunz, F. Califano, Astrophys. J. **856**, L13 (2018)
C.Z. Cheng, G. Knorr, J. Comput. Phys. **22**, 330, (1976)
C.G. Darwin, London, Edinburgh Dublin Philos. Mag. J. Sci. **39**, 537 (1920)
K.J. Genestreti et al., J. Geophys. Res. **123**, 1806 (2018)
C.A. González, T.N. Parashar, D. Gomez, W.H. Matthaeus, P. Dmitruk, Phys. Plasmas **26**, 012306 (2019)
D.J. Griffiths, *Introduction to Electrodynamics* (Cambridge University Press, Cambridge, 1962)
D. Grošelj et al., Astrophys. J. **847**, 28 (2017)
E.G. Harris, Il Nuovo Cimento (1955–1965) **23**, 115 (1962)
M. Hesse, J. Birn, M. Kuznetsova, J. Geophys. Res. **106**, 3721 (2001)
R. Horiuchi, T. Sato, Phys. Plasmas **6**, 4565 (1999)
J.D. Jackson, Am. J. Phys. **70**, 917 (2002)
A.N. Kaufman, P.S. Rostler, Phys. Fluids **14**, 446 (1971)
T.B. Krause, A. Apte, P.J. Morrison, Phys. Plasmas **14**, 102112 (2007)
M.E. Mandt, R.E. Denton, J.F. Drake, Geophys. Res. Lett. **21**, 73 (1994)
A. Mangeney, F. Califano, C. Cavazzoni, P. Trávníček, J. Comput. Phys. **179**, 495 (2002)
S. Markidis, G. Lapenta, R. Uddin, Math. Comput. Simulat. **80**, 1509 (2010)
O. Pezzi et al., J. Plasma Phys. **85**(5), 905850506 (2019)
P.L. Pritchett, IEEE T. Plasma Sci. **28**, 1976 (2000)
P.L. Pritchett, J. Geophys. Res. **106**, 3783 (2001)

P. Ricci, G. Lapenta, J.U. Brackbill, J. Comput. Phys. **183**, 117 (2002)
H. Schmitz, R. Grauer, Phys. Plasmas **13**, 092309 (2006b)
H. Schmitz, R. Grauer, J. Comput. Phys. **214**, 738 (2006a)
M.A. Shay, J.F. Drake, B.N. Rogers, R.E. Denton, J. Geophys. Res. **106**, 3759 (2001)
R.B. Torbert et al., Geophys. Res. Lett. **43**, 5918 (2016)
D.A. Uzdensky, R.M. Kulsrud, Phys. Plasmas **13**, 062305 (2006)
A. Vaivads et al., J. Plasma Phys. **82**, 905820501 (2016)
F. Valentini, P. Trávníček, F. Califano, P. Hellinger, A. Mangeney, J. Comput. Phys. **225**, 753 (2007)
E.S. Weibel, Phys. Rev. Lett. **2**, 83 (1959)
A. Zeiler, D. Biskamp, J.F. Drake, B.N. Rogers, M.A. Shay, M. Scholer, J. Geophys. Res. **107**, 1230 (2002)

Chapter 9
Fully Kinetic Vlasov Simulation of Collisionless Magnetic Reconnection

This chapter presents preliminary results of a 2D antiparallel magnetic reconnection simulation performed with the ViDA code introduced in Chap. 8. The observational study presented in Chap. 7 provides evidence of electron scale structuring of the EDR. As it has been discussed in Chap. 7 and especially in Sect. 7.7, there are only few recent PIC simulations suggesting that the electron diffusion layer may be structured or turbulent (Jara-Almonte et al. 2014; Price et al. 2016, 2017; Swisdak et al. 2018) while in several other studies the EDR appears as a laminar region (Shay et al. 2016; Pritchett 2008; Hesse et al. 2016).

Jara-Almonte et al. (2014) point out that one of the approximations that are used but not critically examined in the framework of numerical simulations is the artificially low ratio between the speed of light and the thermal electron speed $c/v_{th,e}$ which is crucial because it corresponds to the scale separation between the electron skin depth and the Debye length ($c/v_{th,e} = d_e/\lambda_D$).

In particular, they show that while a low $c/v_{th,e}$ may not impact the study of ion scale dynamics, it affects the electron scale dynamics because of electrostatic effects at the Debye scale. Indeed, they find that once significant scale separation is introduced the laminar structure of the EDR is broken. They have performed antiparallel 2D magnetic reconnection simulations using a fully kinetic PIC code and report electrostatic structures at Debye scale in the EDR. These structures are dynamic non-linear electron holes that interact with each other producing turbulence in the electron layer. The instabilities that cause the turbulence are in-plane instabilities and they are a combination of electron-electron and ion-ion stream instabilities.

The typical wavelength of the corresponding mode is found to be approximately 30 λ_D. This means that $d_e/\lambda_D = c/v_{th,e}$ has to be at least 30 for the instability to develop. Figure 9.1 shows that this structuring is visible only when the electron scale and the Debye scale are sufficiently separated. In particular, it develops once $c/v_{th,e}$ exceeds 30, as expected. The quantity plotted in Fig. 9.1 is the electrostatic potential.

G. Cozzani, *Microphysics of Magnetic Reconnection in Near-Earth Space*, Springer Theses, https://doi.org/10.1007/978-3-030-56142-0_9

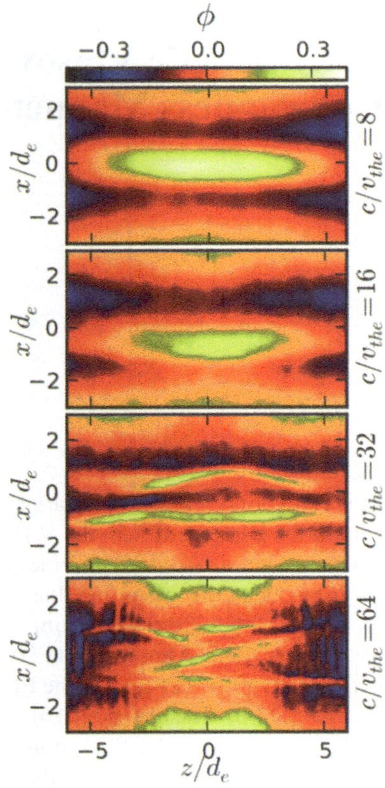

Fig. 9.1 Comparison of the electrostatic potential ϕ obtained in runs with different $c/v_{th,e}$ ratio. In this simulation, the z direction is the outflow direction and the x direction is normal to the current sheet. Reprinted from J. Jara-Almonte, W. Daughton, and H. Ji, Phys. Plasmas 21, 032114 (2014) (Jara-Almonte et al. 2014), with the permission of AIP Publishing

A similar behaviour is observed in other quantities as the out-of-plane current density and the electron pressure.

These recent simulation results, together with the observational evidences of the structured EDR presented in Chap. 7, have provided the motivation for performing a simulation with parameters as close as possible to the runs presented in Jara-Almonte et al. (2014). In particular, I have taken into account run 3 of that study, for which $d_e/\lambda_D = c/v_{th,e} = 32$. The whole simulation setup is discussed in Sect. 9.2. In this chapter, I present results of this 2D antiparallel magnetic reconnection simulation performed with the ViDA code.

In the first part of the chapter, the magnetic reconnection signatures are presented. I anticipate here that the structuring which has been shown by Jara-Almonte et al. (2014) is not found. The reasons for this are still under analysis. Then, the chapter focuses on the evaluation of the contribution of the terms of Ohm's law in the EDR. As already discussed in Sect. 6.1.1, this topic is largely debated. It is found that the electron inertia term, and in particular the term which is proportional to the temporal variation of the current density plays a role in sustaining the reconnection electric field. This contribution is not limited to a region of width 1 d_e around the X-point but it extends for $\sim 5d_p$ in the outflow direction. Of course, this feature can be due to the artificially low mass ratio adopted here.

9.1 Design of the Initial Condition

As it has been already mentioned in Sect. 8.2, the Vlasov-Darwin model implemented in the ViDA code is composed by the Vlasov equation for both electrons and protons and by the Maxwell-Darwin equations written in the form of Helmholtz-like equations (see Eqs. (8.57)–(8.62)). As a consequence, the only quantity that has to be evolved in time is the distribution function for each species. The fields are then evaluated through standard Fast Fourier Transform (FFT) algorithms.

Because of the specific Helmholtz-like form of the electromagnetic set of equations, the initial condition setup must be implemented into the distribution functions instead of in the electromagnetic fields directly. The momenta of the distribution functions are then used as "sources" of the equations of state that allow to compute the fields at each time step.

This section details the initial condition used for the magnetic reconnection runs. As mentioned in Sect. 8.4, it is based on the Harris kinetic equilibrium (Harris 1962) which has been adapted for periodic boundary conditions. Furthermore, a uniform background density is added for both species.

9.1.1 Harris Kinetic Equilibrium

In this subsection we resume the Harris equilibrium (Harris 1962) which is an exact solution of the system of equations

$$\mathbf{v} \cdot \nabla f_\alpha + \frac{Z_\alpha e}{m_\alpha}\left(\mathbf{E} + \frac{\mathbf{v}}{c} \times \mathbf{B}\right) \cdot \nabla_{\mathbf{v}} f_\alpha = 0 \tag{9.1}$$

$$\nabla \cdot \mathbf{E} = 4\pi e \sum_\alpha Z_\alpha \int d\mathbf{v}\, f_\alpha \tag{9.2}$$

$$\nabla \times \mathbf{B} = \frac{4\pi}{c} \sum_\alpha Z_\alpha \int d\mathbf{v}\, \mathbf{v}\, f_\alpha \tag{9.3}$$

where $\alpha = p, e$. Let us assume that all quantities vary in only one direction, namely the y-axis and we take $\mathbf{E} = (0, E_y(y), 0) = (0, -\partial_y \varphi, 0)$ and $\mathbf{B} = (B_x(y), 0, 0)$. In this way, the vector potential $\mathbf{A}$ can be directed along z-axis, $\mathbf{A} = (0, 0, A_z(y))$.

A solution of Eq. (9.1) must be a function of the constants of motion. Here, the energy $W = (m/2)(v_x^2 + v_y^2 + v_z^2) + e\varphi(y)$ and the conjugate momenta to x and z, $p_x = mv_x$ and $p_z = mv_z + \frac{e}{c}A_z(y)$ are the constants of motion. We assume that the distribution functions solution of Eq. (9.1) are Maxwellians with a mean velocity in the z direction named $u_{0,\alpha}$. By inserting Maxwellian distribution functions in Eqs. (9.2)–(9.3) and by writing these expressions in terms of the potentials φ and $\mathbf{A}$ we get

$$\frac{d^2\varphi}{dy^2} = -4\pi n_0 e\left[\exp\left(\frac{e}{T_p c}u_{0,p}A_z - \frac{e}{T_p}\varphi\right) - \exp\left(-\frac{e}{T_e c}u_{0,e}A_z + \frac{e}{T_e}\varphi\right)\right] \tag{9.4}$$
$$\frac{d^2A_z}{dy^2} = -\frac{4\pi n_0 e}{c}\left[u_{0,p}\exp\left(\frac{e}{T_p c}u_{0,p}A_z - \frac{e}{T_p}\varphi\right) - u_{0,e}\exp\left(-\frac{e}{T_e c}u_{0,e}A_z + \frac{e}{T_e}\varphi\right)\right].$$

By imposing

$$\frac{u_{0,p}}{T_p} = -\frac{u_{0,e}}{T_e} \tag{9.5}$$

we find that a solution of Eq. (9.4) is $\varphi = 0$ which implies that $n_p = n_e$ and $\mathbf{E} = 0$ at the initial time. Hence, in order to obtain an expression for $A_z(y)$ we need to solve the following equation

$$\frac{d^2A_z}{dy^2} = -\frac{4\pi n_0 e}{c}(u_{0,p} - u_{0,e})\exp\left(\frac{e}{T_p c}u_{0,p}A_z\right) \tag{9.6}$$

where we made use of the condition of Eq. (9.5). If we take the boundary conditions so that $A_z(y = 0) = 0$, the $A_z(y)$ obtained from Eq. (9.6) reads

$$A_z(y) = \frac{2cT_p}{eu_{0,p}}\log\cosh\left(\frac{u_{0,p}}{T_p}\sqrt{\frac{T_p + T_e}{2}}\sqrt{\frac{4\pi n_0 e^2}{c^2}}y\right). \tag{9.7}$$

Then, if also $B_x(y = 0) = 0$, we obtain the magnetic field profile using the relation $B_x(y) = \partial_y A_z(y)$:

$$B_x(y) = \sqrt{8\pi n_0(T_p + T_e)}\tanh\left(\frac{u_{0,p}}{T_p}\sqrt{\frac{T_p + T_e}{2}}\sqrt{\frac{4\pi n_0 e^2}{c^2}}y\right). \tag{9.8}$$

In the end, the Harris equilibrium provides the following magnetic field

$$B_x(y) = B_0\tanh\left(\frac{y}{L}\right) \tag{9.9}$$

and the following density

$$n_p(y) = n_e(y) = n(y) = \frac{n_0}{\cosh^2\left(\frac{y}{L}\right)} \tag{9.10}$$

and it requires that pressure balance is fulfilled

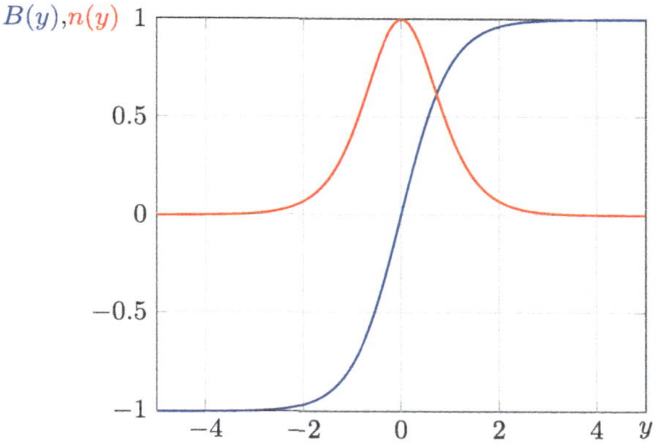

Fig. 9.2 Profiles of the magnetic field and density for the Harris kinetic equilibrium. $L = 1$

$$\frac{B_0^2}{8\pi} = n_0(T_p + T_e). \tag{9.11}$$

Note that B_0 corresponds to the asymptotic absolute value of the magnetic field while n_0 corresponds to the peak of the density in the center of the layer. The magnetic field and density profiles are shown in Fig. 9.2. In addition,

$$L = \frac{2c}{eB_0}\frac{T_p + T_e}{u_{0,p} - u_{0,e}} = -\frac{2c}{eB_0}\frac{T_e}{u_{0,e}} = \frac{2c}{eB_0}\frac{T_p}{u_{0,p}}. \tag{9.12}$$

where the condition of Eq. (9.5) has been used.

9.1.2 *Initial Condition: Double Harris Sheet with GEM-like Perturbation*

The Harris equilibrium discussed in the previous section is needed to present the initial condition designed for a 2D magnetic reconnection simulation with periodic boundary conditions in both x and y direction. In the following, the density profile of Eq. (9.10) is named $n_H(y)$.

The unperturbed distribution functions for electrons f_e and protons f_p read

$$f_e(y, v_x, v_y, v_z) = n_{H,1} \exp\left[-\frac{1}{v_{th,e}^2}[v_x^2 + v_y^2 + (v_z + u_{0,e,1})^2]\right] + \\ + n_{H,2} \exp\left[-\frac{1}{v_{th,e}^2}[v_x^2 + v_y^2 + (v_z - u_{0,e,2})^2]\right] + n_b \exp\left[-\frac{v^2}{v_{th,e}^2}\right] \tag{9.13}$$

$$f_p(y, v_x, v_y, v_z) = n_{H,1} \exp\left[-\frac{1}{v_{th,p}^2} [v_x^2 + v_y^2 + (v_z - u_{0,p,1})^2] \right] +$$
$$+ n_{H,2} \exp\left[-\frac{1}{v_{th,p}^2} [v_x^2 + v_y^2 + (v_z - u_{0,p,2})^2] \right] + n_b \exp\left[-\frac{v^2}{v_{th,p}^2} \right] \qquad (9.14)$$

and they correspond to a double Harris sheet with a background density n_b for both species. The subscript 1 and 2 indicate two different current sheets with different widths L_1 and L_2. The double current sheet configuration is needed because of the periodic boundary conditions. Note that, once the temperature is set, the velocities $u_{0,e,1}$ and $u_{0,p,1}$ (and $u_{0,e,2}$ and $u_{0,p,2}$) are fixed by the current sheet width and the magnetic field asymptotic value (see Eq. (9.12)).

The charge density and the current density are obtained integrating Eqs. (9.13)–(9.14) as follows

$$n_\alpha(y) = e \sum_\alpha Z_\alpha \int d\mathbf{v}\, f_\alpha = e Z_\alpha (n_{H,1} + n_{H,2} + n_b)$$
$$j_{\alpha,x} = e \sum_\alpha Z_\alpha \int d\mathbf{v}\, v_{\alpha,x}\, f_\alpha = 0$$
$$j_{\alpha,y} = e \sum_\alpha Z_\alpha \int d\mathbf{v}\, v_{\alpha,y}\, f_\alpha = 0$$
$$j_{\alpha,z} = e \sum_\alpha Z_\alpha \int d\mathbf{v}\, v_{\alpha,z}\, f_\alpha = e Z_\alpha (n_{H,1} u_{0,\alpha,1} - n_{H,2} u_{0,\alpha,2})$$

Then, the magnetic field is consistently computed using Eq. (8.59). The initial profile of the magnetic field $B_x(y)$, density $n(y)$ and current density $j_z(y)$ are shown in Fig. 9.3. The value of B_0, n_0, L_1 and L_2 are the ones used for the simulation run detailed in Sect. 9.2. The initial temperature profile of both species is uniform.

As already underlined in Sect. 8.4, this is not an exact Vlasov kinetic equilibrium. However, we take care that the initial unperturbed configuration is in force balance and we have checked that it does not undergo significant modifications within the time scale of reconnection considered here.

A perturbation similar to the one used in the so-called GEM challenge (Birn et al. 2001) has been superposed to the initial condition. In particular, the perturbation is meant to obtain an initial single magnetic island and it is assigned through the perturbation on the magnetic flux

$$\delta\psi(x, y) = \psi_0 \cos(2\pi x / L_x) \cos(2\pi y / L_y) \qquad (9.15)$$

where, L_x and L_y are the lengths of the spatial domain in x and y direction respectively. Hence, the fluctuations of the magnetic field $\delta\mathbf{B}$ are given by $\delta\mathbf{B} = \nabla\delta\psi \times \hat{z}$. However, as previously discussed, all the information about the initial condition has

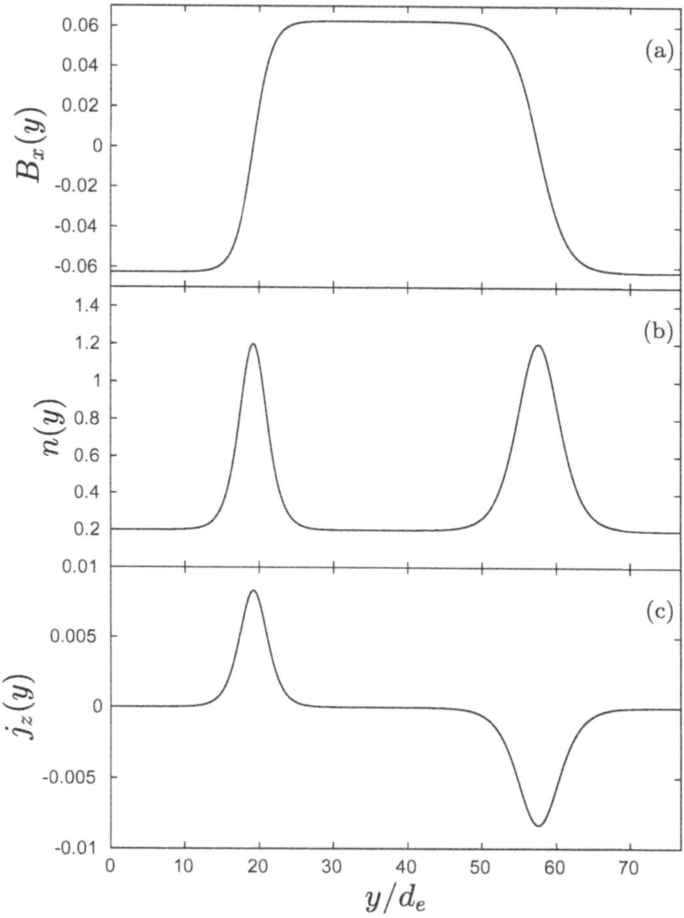

Fig. 9.3 Initial unperturbed profiles of **a** the magnetic field $B_x(y)$; **b** the density $n(y)$ ($n_e = n_p = n$); **c** the current density $j_z(y)$

to be included in the distribution functions. For this reason, a density perturbation $\delta n(x, y)$

$$\delta n(x, y) = -4\pi^2 \left(\frac{1}{L_x^2} + \frac{1}{L_y^2} \right) \delta\psi(x, y) \frac{1}{u_{0,p,1} - u_{0,e,1}}. \tag{9.16}$$

consistent with $\delta\psi(x, y)$ is introduced. Finally, the distribution functions which include the GEM-like perturbations read

$$f_e(y, v_x, v_y, v_z) = (n_{H,1} + \delta n(x, y)) \exp\left[-\frac{1}{v_{th,e}^2}[v_x^2 + v_y^2 + (v_z + u_{0,e,1})^2]\right] + \\ +n_{H,2} \exp\left[-\frac{1}{v_{th,e}^2}[v_x^2 + v_y^2 + (v_z - u_{0,e,2})^2]\right] + n_b \exp\left[-\frac{v^2}{v_{th,e}^2}\right] \quad (9.17)$$

$$f_p(y, v_x, v_y, v_z) = (n_{H,1} + \delta n(x, y)) \exp\left[-\frac{1}{v_{th,p}^2}[v_x^2 + v_y^2 + (v_z - u_{0,p,1})^2]\right] + \\ +n_{H,2} \exp\left[-\frac{1}{v_{th,p}^2}[v_x^2 + v_y^2 + (v_z - u_{0,p,2})^2]\right] + n_b \exp\left[-\frac{v^2}{v_{th,p}^2}\right] \quad (9.18)$$

9.2 Simulation Setup

This section discusses on the parameters that have been used to perform a 2D–3V magnetic reconnection simulation. Reconnection is symmetric and there in no guide field. The *electromagnetic* normalization is adopted (see also Sect. 8.3.3). Lengths are normalized to the electron inertial length $d_e = c/\omega_{p,e}$ at the reference density n_0. Time is normalized to the inverse of electron plasma frequency $\omega_{p,e} = \sqrt{4\pi n_e e^2/m_e}$. Speeds are normalized to the light speed c. The coordinate system is a generic "simulation coordinate system" where the reconnection outflows are directed along the x direction and the inflows are directed along y. Magnetic field strengths and particle number densities are normalized to values B_0 and n_0, respectively. The simulation parameters are listed in Table 9.1.

As mentioned at the beginning of this chapter, the simulation is designed in order to be similar to one of the PIC simulation runs discussed by Jara-Almonte et al. (2014). In particular, I refer to run 3 in Jara-Almonte et al. (2014), see Table II where the simulation box is 100 $d_e \times 50\, d_e$, the mass ratio $m_p/m_e = 25$ and $c/v_{th,e} = 32$. The main feature of the simulations presented in that study is the introduction of significant scale separation between the Debye length and the electron inertial length ($d_e/\lambda_D = c/v_{th,e} = 32$) with all physical scales well resolved, $\Delta x/\lambda_D = 0.8$. For the simulation presented in this chapter it has not been possible to reach the same resolution because of the huge memory and CPU time requirement of an Eulerian code which would require significant computational resources as the one provided for instance in the framework of an EU PRACE computational grant. Nevertheless, as a first step we have performed a simulation using a simulation box 102.4 $d_e \times$ 76.8 d_e wide discretized with $N_x \times N_y = 1024 \times 768$. Hence, the Debye length is not resolved, $\Delta x = 0.1\, d_e = 3.2\, \lambda_D$. The other main difference lies in the boundary conditions which are periodic in the two directions for the simulation presented here

Table 9.1 Simulation input parameters

2D–3V magnetic reconnection run	
L_x/d_e	102.4
L_y/d_e	76.8
N_x	1024
N_y	768
$\Delta x/d_e$	0.1
$\Delta y/d_e$	0.1
m_p/m_e	25
T_p/T_e	1.0
n_b/n_0	0.2
$c/v_{th,e}$	32
L_1/d_e	2.5
L_2/d_e	3.75
B_0	0.0625
n_b/n_0	0.2
T_e/T_p	1.0
$\psi_0/B_0 d_e$	0.1
β	0.2
$\Delta t\ \omega_{p,e}$	0.1

while (Jara-Almonte et al. 2014) simulations have periodic boundary condition in the outflow direction and conducting walls in the direction normal to the current sheet.

Finally, for the simulation presented here the velocity space has been discretized with $N_{e,v_x} \times N_{e,v_y} \times N_{e,v_z} = 41 \times 41 \times 41$ gridpoints in the velocity domain for electrons and $N_{p,v_x} \times N_{p,v_y} \times N_{p,v_z} = 41 \times 41 \times 41$ gridpoints in the velocity domain for protons. Also, $v_e^{max} = 6\ v_{th,e}$ and $v_p^{max} = 8\ v_{th,p}$. The time step is $\Delta t = 0.1\,\omega_{p,e}^{-1}$ which is similar to the time step of $0.14\,\omega_{p,e}^{-1}$ adopted in Jara-Almonte et al. (2014). The width L_1 of the steeper current sheet is $L_1 = 2.5\ d_e = 0.5\ d_p$ and it is consistent with the observational case reported in Chap. 7 for which the current sheet thickness has been measured to be $\sim$0.4 d_i.

9.3 Simulation Results Overview

In this section some of the global properties of the simulation are discussed. Figure 9.4 shows the evolution of the reconnected flux (top panel) and the z component of the electric field normalized to $B_0 V_{A,p}/c$, which is proportional to the reconnection rate (see Sect. 2.2.1). The reconnected flux is given by the difference $\Delta\psi$ between the magnetic flux ψ evaluated at the X-point and at the O-point. Accordingly to the initial

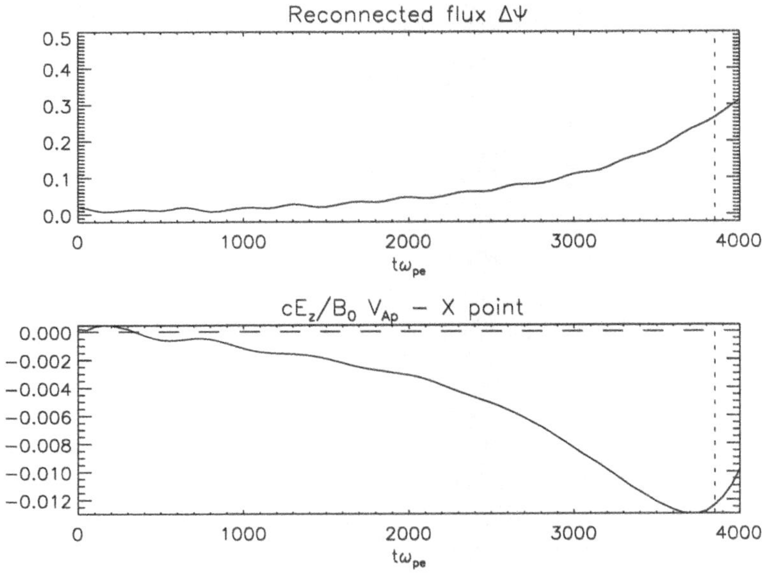

Fig. 9.4 (top) Time evolution of the reconnected magnetic flux $\Delta\psi$; (bottom) time evolution of the normalized reconnection electric field $cE_z/B_0V_{A,p}$ at the X-point

perturbation, the X-point and the O-point are initially located at $(L_x/2, L_y/4)$ and $(0, L_y/4)$ and their location does not significantly change throughout the simulation run.

The absolute value of normalized reconnection electric field increases until it reaches its maximum value in the interval $t = [3650, 3800]\,\omega_{p,e}^{-1}$ $(\Omega_{c,p}^{-1} = 400\,\omega_{p,e}^{-1})$. In that interval, reconnection is ongoing and the reconnection rate is constant. Immediately after, E_z (and the reconnection rate) starts to decrease. The vertical dotted line indicates the time $(t = 3852\,\omega_{p,e}^{-1})$ at which an island is formed where the X-point was standing. The newly formed O-point does not move and there are two newly formed X-point which are symmetric relatively to the O-point. In the following, the analysis will be performed on the time period previous to the formation of the island $(t < 3852\,\omega_{p,e}^{-1})$ and in particular in the interval in which E_z is constant.

Note that the evolution of the reconnected flux and of E_z is affected on long times by the periodic boundary conditions which do not allow the system to reach a steady state condition for which E_z and the reconnected flux are stable for several tens of characteristic proton times $\Omega_{c,p}^{-1}$. The period of time in which the reconnection rate can be considered as constant is limited to the peak of the electric field and it lasts for about $150\,\omega_{p,e}^{-1}$. The following analysis is performed within this time interval.

Figures 9.5 and 9.6 are meant to give an overview of the main characteristics of magnetic reconnection. Figure 9.5b shows that the out-of-plane magnetic field exhibits the typical quadrupolar pattern (see Sect. 2.4) while the reconnection electric field E_z is concentrated around the X-point. The electron and ion outflow are shown

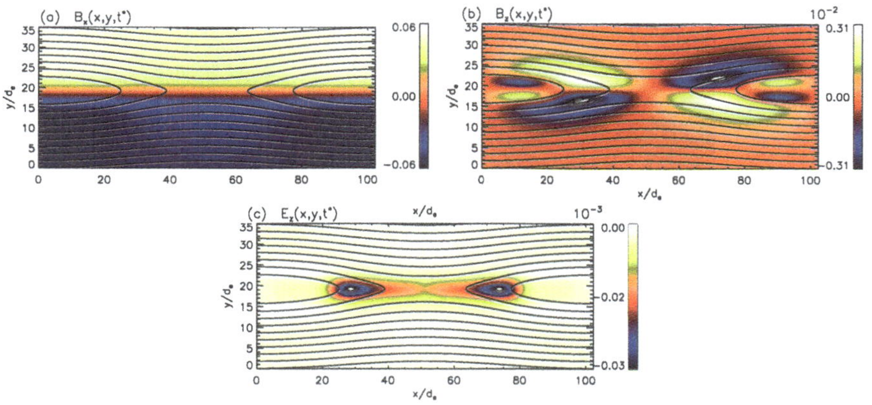

Fig. 9.5 Contour plot of **a** the reconnecting magnetic field component B_x; **b** the out-of-plane magnetic field B_z; **c** the reconnection electric field E_z. All quantities are shown at time $t^* = 3752\ \omega_{p,e}^{-1}$ and zoomed in y in the interval $[0, 35]\ d_e$. The contour lines of the magnetic flux ψ are superposed

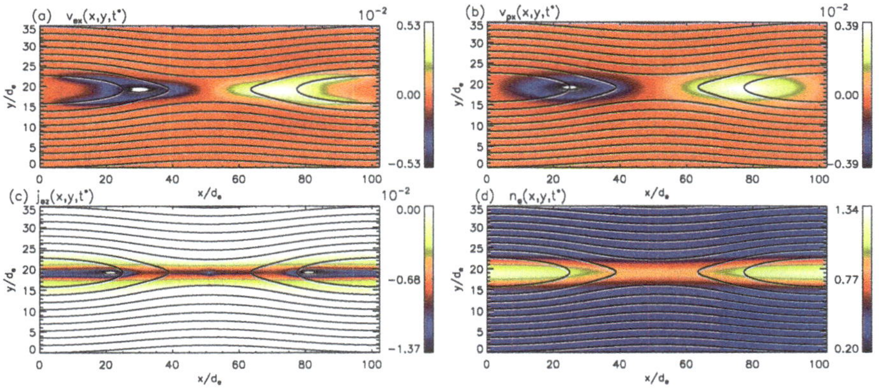

Fig. 9.6 Contour plot of **a** the electron outflow velocity $v_{e,x}$; **b** the proton outflow velocity $v_{p,x}$; **c** the out-of-plane electron current density $j_{e,z}$; **d** the electron density n_e. All quantities are shown at time $t^* = 3752\ \omega_{p,e}^{-1}$ and zoomed in y in the interval $[0, 35]\ d_e$. The contour lines of the magnetic flux ψ are superposed

in Fig. 9.6a, b. The electron outflow turns out to be bifurcated along the separatrices while the ion outflow is concentrated in the center of the outflow region. The width of the out of plane current $j_{z,e}$ in the y direction is about 2.5 d_e, so the current sheet has not become much thinner during the evolution with respect to the initial condition (Fig. 9.6c). However, the current sheet is slightly thinner at the X-point. Also, the density is depleted at the X-point while it increases within the magnetic island (see Fig. 9.6d).

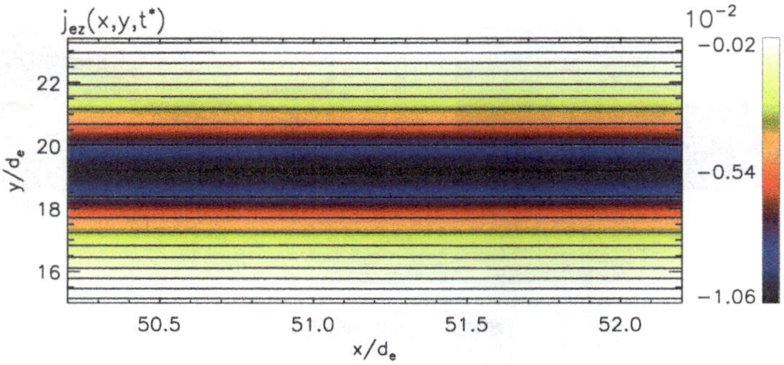

Fig. 9.7 Contour plot of the out-of-plane electron current density at time $t^* = 3752\ \omega_{p,e}^{-1}$ and zoomed in y in the interval $[15.0, 23.4]\ d_e$ and in x in the interval $[50.2, 52.2]\ d_e$. The contour lines of the magnetic flux ψ are superposed

As mentioned at the beginning of the chapter, the choice of parameters for this simulation was motivated by the simulations performed by Jara-Almonte et al. (2014) which, as already discussed, present results that are consistent with the observational case reported in Chap. 7. Indeed, the simulations performed by Jara-Almonte et al. (2014) show structures at and below the electron scale in the EDR.

However, these structures are not present in the simulation run presented here. We can see that the different quantities are rather laminar, in agreement with other simulations where the Debye length is not resolved as well (Shay et al. 2007) but that have a lower $c/v_{th,e}$ ratio of about 10. Figure 9.7 shows in particular a zoom of the out-of-plane electron current density in an interval 2 d_e long in the x-direction (the X-point is located at $x = 51.2\ d_e$). We see that $j_{e,z}$ is not exhibiting structures or signatures of turbulence. The reasons for this discrepancy are currently unknown. As discussed in Sect. 9.2, the simulation presented here as a lower resolution than the one discussed in Jara-Almonte et al. (2014). However, being the typical wavelength of the expected modes of the order of $30\ \lambda_D \sim 9.4\ \Delta x$ ($\Delta_x = 3.2\ \lambda_D$), the resolution of this run should be high enough to show at least some deviations from laminarity, if these deviations were present. Therefore, it is not possible to draw a definite conclusion. The investigation of this discrepancy will be part of future work.

9.4 Electron Dynamics in the Current Layer

The motion of the electrons in a plasma is described by the fluid momentum equation for electrons

$$\mathbf{E} = -\frac{\mathbf{u}_e \times \mathbf{B}}{c} - \frac{1}{n_e e}\nabla \cdot \overline{\mathbf{P}_e} + \frac{m_e}{n_e e^2}\frac{\partial}{\partial t}(\mathbf{j}_e) - \frac{m_e}{n_e e}\nabla \cdot (n\mathbf{u}_e\mathbf{u}_e) \tag{9.19}$$

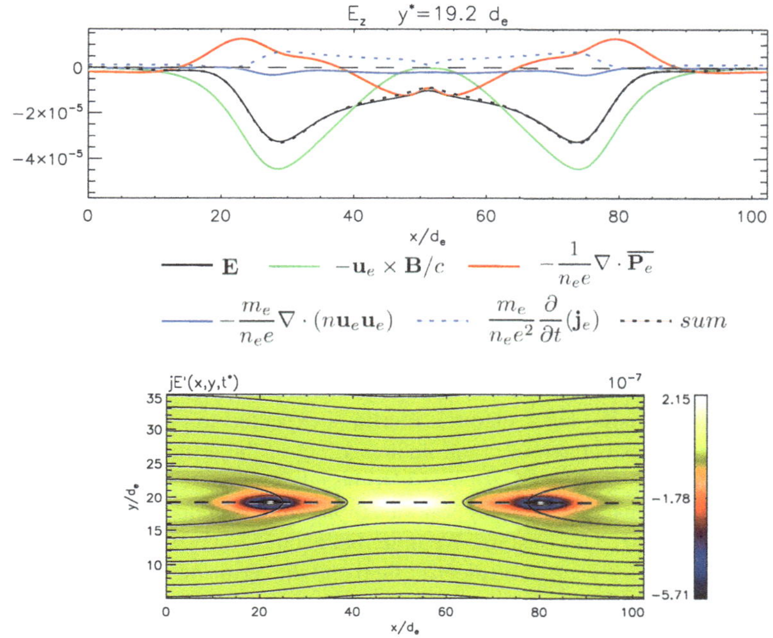

Fig. 9.8 (top) Cuts through the X-point of the contributions to Ohm's law in the z direction at time $t^* = 3727\ \omega_{p,e}^{-1}$; (bottom) contour plot of the energy conversion $\mathbf{E}' \cdot \mathbf{j}\ t^* = 3727\ \omega_{p,e}^{-1}$ zoomed in y in the interval $[0, 35]\ d_e$. The contour lines of the magnetic flux ψ are superposed

where $\mathbf{j}_e = -en_e\mathbf{u}_e$. In this section, we discuss the contribution of the different terms on the right hand side of Eq. (9.19) in composing the reconnection electric field E_z. Equation (9.19) is also referred to as "Ohm's law" since it is written is a Ohm's law fashion

$$\mathbf{E} = -\frac{\mathbf{u}_e \times \mathbf{B}}{c} + \mathbf{R}$$

where $\mathbf{R}$ contains all the non-ideal terms.

The analysis is performed at the time corresponding to the E_z peak, as previously discussed.

Figure 9.8(top) shows the behavior of the terms in Eq. (9.19) along a cut through the X-point along the outflow direction. The cut corresponds to the black dashed line in Fig. 9.8(bottom). The sum of all terms (black dotted line) agrees with the electric field E_z (black line).

Concerning the relative role of the different terms, one can see that while the ideal term $(\mathbf{u}_e \times \mathbf{B})_z$ goes to zero close to the X-point (light green line), the term involving the divergence of $\overline{\mathbf{P}_e}$ (pressure term, red line) becomes larger and it basically sustains the electric field in the inner diffusion region. The boundary of the inner electron diffusion region can be identified with the point where the Lorentz force $(\mathbf{u}_e \times \mathbf{B})_z$

crosses the reconnection electric field E_z (Shay et al. 2007). The inner EDR can also be defined in terms of the energy dissipation $\mathbf{E}' \cdot \mathbf{j}$ (Fig. 9.8,(bottom)). In particular, $\mathbf{E}' \cdot \mathbf{j} > 0$ identifies the inner EDR and $\mathbf{E}' \cdot \mathbf{j} < 0$ the outer EDR (Karimabadi et al. 2007) (see also Sect. 6.1.2, $\mathbf{E}' = \mathbf{E} + \mathbf{u}_e \times \mathbf{B}$). Using both methods, the inner diffusion region is found to extend in the outflow for 12.7 d_e (2.54 d_p) from the X-point, which is consistent with previous studies with the same mass ratio (Shay et al. 2007).

The electron inertia term is composed by a term involving the divergence of the velocity and another proportional to the temporal variation of the current (light blue solid and dashed lines). It is found to be non negligible both in the inner and outer EDR. In particular, the term $\propto \partial \mathbf{j}_e / \partial t$ appears to be larger than the $\nabla \cdot (n_e \mathbf{u}_e \mathbf{u}_e)$ term. Furthermore, these terms are not only important in close proximity to the X-point but they rather extends for several d_e along the outflow direction. Both contributions to electron inertia become zero only at a distance of about 24 d_e ($\sim$5 d_p) from the X-line. This is quite surprising since electron inertia is expected to have a role at electron scales; however this behaviour could be an artifact of the artificially low mass ratio of 25. The fact the term $\propto \partial \mathbf{j}_e / \partial t$ needs to be taken into account for Eq. (9.19) to be fulfilled indicates that the temporal variations may be important even when the reconnection rate (E_z) is constant. We note that even if Fig. 9.8 is made at a specific time instant, $t^* = 3727\, \omega_{p,e}^{-1}$, the electric field and the terms on the right hand side of Eq. (9.19) do not vary significantly during the whole interval $t = [3650, 3800]\, \omega_{p,e}^{-1}$.

9.5 Discussion and Conclusions

In this chapter, a preliminary analysis of a 2D antiparallel magnetic reconnection simulation performed with the full kinetic ViDA code is discussed. As detailed at the beginning of the chapter, the input parameters of the simulation were chosen to be as close as possible to one of the runs discussed by Jara-Almonte et al. (2014) which presents electron and Debye scale structuring of the current layer and the development of turbulence that could be consistent with the magnetopause observations presented in Chap. 7. However, in our simulation the electron diffusion region turns out to be rather laminar. As underlined in Sect. 9.3, even though the resolution of the run presented here is lower than the one adopted in the corresponding run in Jara-Almonte et al. (2014), it should be enough to at least hint possible deviations from laminarity. At present, this discrepancy has not been further investigated but the reasons behind it should be clarified in future work. Since the two main differences between the two runs are the resolution and the boundary conditions (see Sect. 9.2) this investigation would benefit from reaching the same resolution of run 3 as in Jara-Almonte et al. (2014).

The chapter is then devoted to the analysis of the different terms composing the reconnection electric field. In particular, in agreement with previous studies (Shay et al. 2007; Divin et al. 2012, 2016) it is found that in the inner diffusion region the reconnection electric field is mainly sustained by the divergence of the electron pressure tensor, in particular by the out-of-diagonal elements. However, differently from previous studies of antiparallel magnetic reconnection, the inertia term is found to be non negligible and to extend for several d_e in the outflow direction. The main contribution to the inertia term is given by the temporal variation of the electron current density. However, the fact that the inertia term is important for about 24 d_e in the outflow direction can be due to the artificially low mass ratio ($m_p/m_e = 25$) that has been adopted for this run. Indeed, Shay et al. (2007) performed several antiparallel magnetic reconnection runs with increasing mass ratio ($m_p/m_e = 25, 100, 400$) showing that the length of the electron current layer becomes significantly shorter as the mass ratio is increased. Similar findings are presented by Le et al. (2013). Hence, we could expect that the region in which the electron inertia term is playing a role would be actually reduced if a realistic mass ratio is adopted.

The interesting finding is that the main contribution to the inertia term is given by the temporal variation of the electron current density $\partial \mathbf{j}_e/\partial t$. This term is usually not included in the evaluation of the Ohm's law in other studies (Shay et al. 2007; Divin et al. 2012, 2016, see e.g.) or it is found to be small compared to the other terms everywhere (Hesse et al. 2014, 2016). In addition, it is extremely difficult to evaluate this term with spacecraft data and this result can be useful to interpret the residue between the electric field and the right hand side of Ohm's law which is often found in observations (Torbert et al. 2016a), where $\partial \mathbf{j}_e/\partial t$ is usually not computed.

However, in order to compare in a meaningful way our result with previous findings, we need to take into account the specific time at which the analysis was performed in terms of evolution of the magnetic reconnection process. For instance, the analysis by Shay et al. (2007) on a run with mass ratio $m_p/m_e = 25$ is performed when the reconnection rate remains stable for several $\Omega_{c,p}^{-1}$. This condition is never achieved in the simulation presented here, as detailed in Sect. 9.3. This is probably due to the periodic boundary conditions that limit the time interval in which the simulation data are physically meaningful (Daughton et al. 2006) combined with the rather limited size of the box. A larger simulation box is needed to let the system reach a proper steady state for a sufficiently long time interval.

In the end, we argue whether it is meaningful to make such a comparison and we rather underline that the analysis presented here is performed in another phase of the reconnection process, namely when the reconnection electric field reaches its peak (together with the reconnection rate) but the reconnection process is not in a "traditional" steady state. In this phase, in which reconnection is rather unsteady, the current density variation in time appears to be a non negligible contribution which is needed to fulfill Eq. (9.19). In summary, we think that in order to reach a definite conclusion it is necessary to perform other runs (e.g. with a larger simulation box) in order to clarify the role of the periodicity of the boundary conditions. On the other side, quantifying the role of term proportional to $\partial \mathbf{j}_e/\partial t$ in sustaining the reconnection electric field in the unsteady phase can provide insights about the electron energization in this phase of the reconnection process and it would allow to establish how much of the energization is due to the time varying term of Ohm's law which are not considered in the steady state studies. Future work must include also the analysis of the electrons distribution functions.

9.6 Future Work

This section provides an outlook of possible follow-up studies of reconnection simulations performed with the full kinetic ViDA code. There are several future work planned based on this new code – and they are discussed in Chap. 10 – but in this section I present a different initial perturbation for magnetic reconnection simulations that represents an alternative to the GEM-like perturbation. Indeed, in most of the recent numerical studies, magnetic reconnection is initiated by imposing a single X-point in the current sheet (Shay et al. 2007; Divin et al. 2016, e.g.). This can be done either with a domain-large perturbation present in all the simulation box (Birn et al. 2001) or with a perturbation which is localized around the location where reconnection should start and which has a initially imposed wavelength (Lapenta et al. 2010; Divin et al. 2012). This initial condition has been used in the context of the GEM challenge to willingly put the system in the nonlinear phase of magnetic reconnection from the beginning of the simulation. As a consequence, while these single X-point perturbations allow to successfully study the steady state of reconnection, they do not allow to investigate in a self-consistent way the growth of the tearing instability (Swift 1986), the formation and the characteristics of plasmoid chains (Loureiro et al. 2007; Markidis et al. 2012, 2013) or the long standing problem of the reconnection onset.

For this reason, this section presents an initial condition for a 2D magnetic reconnection simulation composed by the double Harris sheet discussed in Sect. 9.1 to which we add a small amplitude perturbation composed by fluctuations with multiple in-plane wave vectors $k_x = 2\pi n/L_x$ and $k_y = 2\pi m/L_y$ (in this case $n, m = 0, \ldots, 9$ and $\sqrt{n^2 + m^2} \leq 9$) with random phases. This section is limited to the description of the initial condition and to present preliminary results in which it can be seen that reconnection is taking place. The analysis of the simulation data will be carried on in the future, together with other runs with higher resolution. The final long term goal is to study the mechanisms at play as the onset of the reconnection process and the dynamics of plasmoid chains with the fully kinetic Vlasov code ViDA.

9.6.1 Initial Condition

Fluctuations are defined by means of the vector potential $\delta\mathbf{A}(x, y)$ in order to ensure the solenoidality of the magnetic field perturbation $\delta\mathbf{B}(x, y)$.

$$\delta A_x(x, y) =$$

$$= \epsilon \sum_{r=1}^{N+1} \sum_{n=0}^{N} \sum_{m=1}^{M} \frac{1}{k_{y,m}} [\cos(k_{x,n}x + k_{y,m}y + \phi_{x,r}) + \cos(k_{x,n}x - k_{y,m}y + \psi_{x,r})]$$

$$\delta A_y(x, y) =$$

$$= \epsilon \sum_{r=1}^{N+1} \sum_{n=1}^{N} \sum_{m=0}^{M} \frac{1}{k_{x,n}} [\cos(k_{x,n}x + k_{y,m}y + \phi_{y,r}) + \cos(k_{x,n}x - k_{y,m}y + \psi_{y,r})]$$

$$\delta A_z(x,y) = \\ = \epsilon \sum_{r=1}^{N+1} \sum_{n=0}^{N} \sum_{m=0}^{M} \frac{1}{\sqrt{k_{x,n}^2 + k_{y,m}^2}} [\cos(k_{x,n}x + k_{y,m}y + \phi_{z,r}) + \\ + \cos(k_{x,n}x - k_{y,m}y + \psi_{z,r})] \tag{9.20}$$

where $\phi_{i,r}$ and $\psi_{i,r}$ $(i = x, y, z)$ are random phases and ϵ is the amplitude of the fluctuations.

The magnetic field fluctuations $\delta\mathbf{B}(x, y)$ are obtained using $\delta\mathbf{B}(x, y) = \nabla \times \delta\mathbf{A}(x, y)$. Since all the information about the initial condition has to be included in the distribution functions, as previously discussed in Sect. 9.1, the current density fluctuations are defined consistently as

$$\delta\mathbf{j}(x, y) = \nabla \times \delta\mathbf{B}(x, y)$$

together with the corresponding velocity fluctuations

$$\delta\mathbf{v}(x, y) = \frac{\delta\mathbf{j}(x, y)}{n_b}$$

where n_b is a uniform background density. In the end, analogously to Sect. 9.1, the distribution functions including the random fluctuations $\delta\mathbf{v}(x, y)$ read

$$f_e(y, v_x, v_y, v_z) = n_{H,1} \exp\left[-\frac{1}{v_{th,e}^2}[v_x^2 + v_y^2 + (v_z + u_{0,e,1})^2]\right] + \\ + n_{H,2} \exp\left[-\frac{1}{v_{th,e}^2}[v_x^2 + v_y^2 + (v_z - u_{0,e,2})^2]\right] + n_b \exp\left[-\frac{(\mathbf{v} + \delta\mathbf{v}/2)^2}{v_{th,e}^2}\right] \tag{9.21}$$

$$f_p(y, v_x, v_y, v_z) = n_{H,1} \exp\left[-\frac{1}{v_{th,p}^2}[v_x^2 + v_y^2 + (v_z - u_{0,p,1})^2]\right] + \\ + n_{H,2} \exp\left[-\frac{1}{v_{th,p}^2}[v_x^2 + v_y^2 + (v_z - u_{0,p,2})^2]\right] + n_b \exp\left[-\frac{(\mathbf{v} - \delta\mathbf{v}/2)^2}{v_{th,p}^2}\right] \tag{9.22}$$

Note that in this case the perturbation is present in all the simulation domain and that both proton and electron distribution functions are perturbed. However, this can be modified, and the fluctuations could also be set to be localized only in the region of the Harris sheet.

Table 9.2 Simulation input parameters

2D–3V magnetic reconnection run	
L_x/d_e	102.4
L_y/d_e	76.8
N_x	512
N_y	384
$\Delta x/d_e$	0.2
$\Delta y/d_e$	0.2
m_p/m_e	25
T_p/T_e	1.0
n_b/n_0	0.2
$c/v_{th,e}$	32
L_1/d_e	2.5
L_2/d_e	3.75
B_0	0.0625
n_b/n_0	0.2
T_e/T_p	1.0
β	0.2
$\Delta t\ \omega_{p,e}$	0.1
ϵ	0.001

9.6.2 Simulation Setup and Overview

A 2D–3V magnetic reconnection simulation has been performed to test the initial conditions with random fluctuations presented above. Reconnection is symmetric and there in no guide field. The *electromagnetic* normalization is adopted (see also Sect. 8.3.3). Lengths are normalized to the electron inertial length $d_e = c/\omega_{p,e}$ at the reference density n_0. Time is normalized to the inverse of electron plasma frequency $\omega_{p,e} = \sqrt{4\pi n_e e^2/m_e}$. Speeds are normalized to the light speed c. Magnetic field strengths and particle number densities are normalized to values B_0 and n_0, respectively. The simulation parameters are listed in Table 9.2.

Finally, the velocity space has been discretized with $N_{e,v_x} \times N_{e,v_y} \times N_{e,v_z} = 41 \times 41 \times 41$ gridpoints in the velocity domain for electrons and $N_{p,v_x} \times N_{p,v_y} \times N_{p,v_z} = 41 \times 41 \times 41$ gridpoints in the velocity domain for protons. Also, $v_e^{max} = 6\ v_{th,e}$ and $v_p^{max} = 8\ v_{th,p}$. The time step is $\Delta t = 0.1\ \omega_{p,e}^{-1}$. The width L_1 of the steeper current sheet is $L_1 = 2.5\ d_e = 0.5\ d_p$ while $L_2 = 3.75\ d_e$.

Figure 9.9 shows the evolution of the magnetic flux ψ for the steeper current sheet with width L_1. One can see (Fig. 9.9a) that at the beginning the current sheet is perturbed with small amplitude random fluctuations; then, multiple X and O-points are generated (Fig. 9.9b). At later times the islands tend to merge and in the end a single X-point is left (Fig. 9.9d).

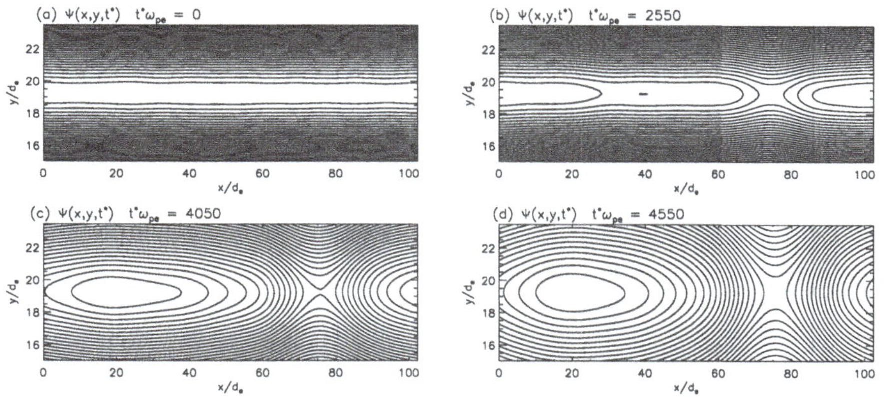

Fig. 9.9 Contour plot of the magnetic flux ψ at different times **a** $t^* = 0\ \omega_{p,e}^{-1}$; **b** $t^* = 2550\ \omega_{p,e}^{-1}$; **c** $t^* = 4050\ \omega_{p,e}^{-1}$; **d** $t^* = 4550\ \omega_{p,e}^{-1}$. All quantities are zoomed in the interval $y = [15, 23.4]\ d_e$

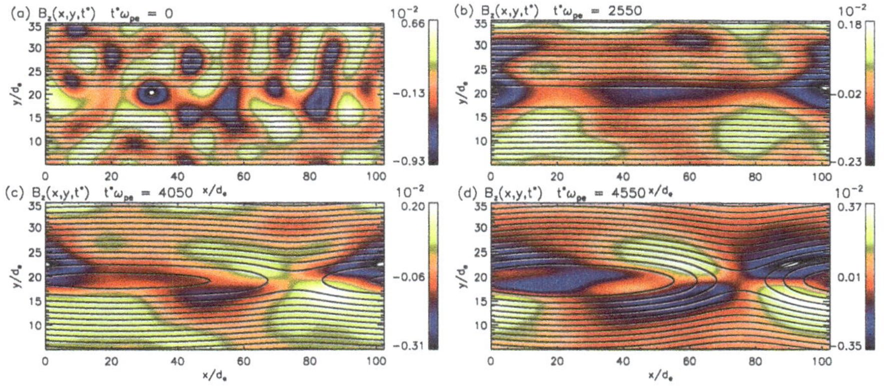

Fig. 9.10 Contour plot of the out-of-plane magnetic field B_z at different times **a** $t^* = 0\ \omega_{p,e}^{-1}$; **b** $t^* = 2550\ \omega_{p,e}^{-1}$; **c** $t^* = 4050\ \omega_{p,e}^{-1}$; **d** $t^* = 4550\ \omega_{p,e}^{-1}$. All quantities are zoomed in y in the interval $[5, 35]\ d_e$. The contour lines of the magnetic flux ψ are superposed

Figure 9.10 shows the evolution of the out-of-plane magnetic field B_z and we can see that the single X-point left presents a quadrupolar magnetic field pattern, a clear signature of ongoing reconnection. In conclusion, magnetic reconnection is successfully reproduced using an initial condition with random perturbations rather then the single X-point GEM perturbation. At the best of my knowledge, this kind of simulations is original for current sheets of characteristic width less then d_p; it could be used to investigate topics and phase of the reconnection process that can not be studied with a GEM-like perturbation of the current sheet, as discussed at the beginning of this section.

References

J. Birn et al., J. Geophys. Res. **106**, 3715 (2001)

W. Daughton, J. Scudder, H. Karimabadi, Phys. Plasmas **13**, 072101 (2006)

A. Divin, G. Lapenta, S. Markidis, V.S. Semenov, N.V. Erkaev, D.B. Korovinskiy, H.K. Biernat, J. Geophys. Res. **117**, A06217 (2012)

A. Divin, V. Semenov, D. Korovinskiy, S. Markidis, J. Deca, V. Olshevsky, G. Lapenta, Geophys. Res. Lett. **43**, 10565 (2016)

E.G. Harris, Il Nuovo Cimento (1955–1965) **23**, 115 (1962)

M. Hesse, N. Aunai, D. Sibeck, J. Birn, Geophys. Res. Lett. **41**, 8673 (2014)

M. Hesse, Y.-H. Liu, L.-J. Chen, N. Bessho, M. Kuznetsova, J. Birn, J.L. Burch, Geophys. Res. Lett. **43**, 2359 (2016)

J. Jara-Almonte, W. Daughton, H. Ji, Phys. Plasmas **21**, 032114 (2014)

H. Karimabadi, W. Daughton, J. Scudder, Geophys. Res. Lett. **34**, L13104 (2007)

G. Lapenta, S. Markidis, A. Divin, M. Goldman, D. Newman, Phys. Plasmas **17**, 082106 (2010)

A. Le, J. Egedal, O. Ohia, W. Daughton, H. Karimabadi, V. S. Lukin, Phys. Rev. Lett. **110**, 135004 (2013)

N.F. Loureiro, A.A. Schekochihin, S.C. Cowley, Phys. Plasmas **14**, 100703 (2007)

S. Markidis, P. Henri, G. Lapenta, A. Divin, M.V. Goldman, D. Newman, S. Eriksson, Nonlin. Processes Geophys. **19**, 145 (2012)

S. Markidis, P. Henri, G. Lapenta, A. Divin, M.V. Goldman, D. Newman, E. Laure, Phys. Plasmas **20**, 082105 (2013)

L. Price, M. Swisdak, J.F. Drake, P.A. Cassak, J.T. Dahlin, R.E. Ergun, Geophys. Res. Lett. **43**, 6020 (2016)

L. Price, M. Swisdak, J.F. Drake, J.L. Burch, P.A. Cassak, R.E. Ergun, J. Geophys. Res. **122**, 11086 (2017)

P.L. Pritchett, J. Geophys. Res. **113**, A06210 (2008)

M.A. Shay, J.F. Drake, M. Swisdak, Phys. Rev. Lett. **99**, 155002 (2007)

M.A. Shay et al., Geophys. Res. Lett. **43**, 4145 (2016)

D.W. Swift, J. Geophys. Res. **91**, 219 (1986)

M. Swisdak, J.F. Drake, L. Price, J.L. Burch, P.A. Cassak, T.D. Phan, Geophys. Res. Lett. **45**, 5260 (2018)

R.B. Torbert et al., Geophys. Res. Lett. **43**, 5918 (2016)

Chapter 10
Conclusions and Outlook

The work presented in this Thesis focus on the microphysics of magnetic reconnection and in particular on the structure and the dynamics of the Electron Diffusion Region (EDR). This topic has been investigated by using two different but complementary approaches, namely in situ data observations from the MMS mission and Eulerian kinetic Vlasov numerical simulations.

As discussed in Chaps. 3, 5 and 6, magnetic reconnection in the Earth's magnetosphere has been extensively studied for about 60 years now. Important steps towards a complete understanding of the process from the largest to smallest scale has been made by means of spacecraft observations of THEMIS, Cluster and MMS in synergy with fluid, kinetic-hybrid (ion kinetic, fluid electrons) and fully kinetic simulations. On the other hand, fundamental questions about the physics at the electron scale remain unsolved and only recently, thanks to high resolution particle measurements in the magnetosphere and to the increasing computational power of supercomputers, the electron scale processes at play in the diffusion region have started to be enlightened.

In this Thesis, I tried to reply to one of the key questions that are still open in the context of the EDR, which is whether the EDR has a preferred homogeneous or inhomogeneous structure at electron scales and below in terms of current densities and electric fields. A non laminar structure would affect the energy conversion in the diffusion region and it could affect the magnetic reconnection process at large scales.

The observational part of the project started with an extended search of MMS sets of data (events) suited for studying the structure of the EDR at the Earth's magnetopause, both in terms of spacecraft orbit and inter-spacecraft separation. Then, I selected one EDR event with the smallest inter-spacecraft separation (~6 km) which has allowed to perform a multi-spacecraft study highlighting the differences in field and particle signatures at electron scales using data from the flux-gate magnetometer, 3D electric field data from the axial and spin-plane probes and particles data from

G. Cozzani, *Microphysics of Magnetic Reconnection in Near-Earth Space*, Springer Theses, https://doi.org/10.1007/978-3-030-56142-0_10

the fast plasma instrument. In particular, I have carefully analyzed MMS electric field data and I have treated in detail the associated experimental errors. This error analysis was crucial to quantitatively assess the energy conversion in the EDR, which is typically small and associated to large errors. The results of this observational study provided evidence that the structure of the EDR at the magnetopause can be much more complex than it has been found in previous MMS events and than it is usually depicted by PIC simulations. In particular, this study shows that EDR is inhomogeneous and that the pattern of the energy conversion is patchy, suggesting a possible turbulent evolution of the EDR structure during reconnection.

The MMS data analysis have pointed out the need of simulations with high spatial resolution and low noise that could be used in synergy with observations in order to understand the kinetic physics at play at electron scales. Following this motivation, the second part of research work has been tailored to study the EDR by using a fully-kinetic Eulerian algorithm solving the Vlasov-Maxwell system of equations in the so-called Darwin approximation (the ViDA code, Chap. 8). The ViDA code is specifically designed to improve our understanding of the kinetic dynamics of collisionless plasma at electron scales by giving access to the fine phase space details of the electron distribution function. This numerical tool is specifically suited for the study of the physical processes in the EDR and other electron-scale reconnection regions such as separatrices and jet fronts and, in particular, for the processes responsible for the patchy structure of these regions, for turbulent anomalous resistivity, for intermittent heating and electrons acceleration. The ViDA code has been developed in the last year in the framework of a collaboration between Università di Pisa and Università della Calabria. I have contributed to the testing of this new code for which I have been responsible, in particular, for simulating magnetic reconnection in a parameter range of interest for satellite observations. The results of the 2D symmetric magnetic reconnection simulations have been presented in Chap. 8 where it is shown that the code correctly reproduces the main properties of magnetic reconnection.

The next step has been to perform 2D symmetric magnetic reconnection simulations with higher resolution compared to the one presented in Chap. 8 and to investigate the contribution of the different terms composing the reconnection electric field in the electrons equation of motion. This problem can be properly addressed by the ViDA code due to the very low computational noise of electromagnetic fields and particle distributions. The preliminary analysis reported in Chap. 9 suggests that the electron inertia term, and in particular the portion including the temporal derivative of the current density, could actually play a role in sustaining the reconnection electric field. However, as detailed in Sect. 9.5, the effects that may be played by the presence of periodic boundary conditions and a relatively small simulation box have to be carefully taken into account before drawing conclusions. New simulation runs performed with larger boxes will improve our understanding of the role of the inertia term in the different phases of the reconnection process.

The research work presented here has tried to reply to few, but fundamental questions but it has also brought out new questions to be answered and new paths of investigation.

Although the observational MMS study presented here has been mainly devoted to magnetic reconnection at the magnetopause, I have devoted some time to the analysis of magnetotail MMS data. I have selected a magnetotail EDR events in order to investigate the structure of the EDR, similarly to what has been done in Chap. 7 for the magnetopause EDR. The magnetotail event has been preliminary presented in Sect. 7.8. Observations of energy conversion during this EDR encounter suggest that the energy conversion can be patchy also at the magnetotail EDR. I would like to continue this line of research trying to clarify whether and to which extent the inhomogeneities in the EDR at the magnetotail are present and have the same origin as the ones at the magnetopause.

A longer term outlook would be to study the structure of the EDR statistically and try to understand which conditions (local and large scale) induce the structuring of the EDR or its laminarity.

As detailed in Chap. 9, the input parameters of the simulation presented there were chosen to be as close as possible to one of the runs discussed by Jara-Almonte et al. (2014) which presents electron and Debye scale structuring of the current layer and the development of turbulence that could be consistent with the magnetopause observations presented in Chap. 7. However, the electron diffusion region emerging from the simulation discussed in Chap. 9 seems to be rather laminar. In order to clarify which are the reasons behind this difference, I would like to perform high resolution simulations such as those run by Jara-Almonte et al. (2014). As already discussed in Sect. 9.2, for the simulation presented in this Thesis, it has not been possible to reach the same resolution because of the huge memory and CPU time requirement of an Eulerian code which would require significant computational resources accessible only in the framework of very large computational grants, such as those within the PRACE EU program.

Also, the ViDA code is a highly valuable tool to study electron scale physics in the frame of magnetic reconnection but also the turbulence cascade below sub-proton scales, which still has to be properly exploited. ViDA is a newborn code that I still would like to use for several future work. In particular, future work could include

1. performing 3D simulations of magnetic reconnection. In this Thesis, only 2D simulations are presented but the code has been designed to allow for 3D runs which, of course, would provide data closer to reality.
2. performing magnetic reconnection simulations adopting initial conditions different from the GEM-like perturbation (Birn et al. 2001). As discussed in Sect. 9.6, the GEM-like initial condition does not allow to study the development of the instability since a single X-point is imposed by the initial perturbation. A different initial condition which could allow to study the early stage of magnetic reconnection has been already designed and tested (see Sect. 9.6.1). The future work would include the investigation of the role of the electron physics in the early stage of magnetic reconnection by means of this kind of approach.
3. adapting the symmetric initial condition to the asymmetric case in order to study reconnection in a magnetopause-like configuration. Recently, Allanson et al. (2017) presented a new and exact equilibrium solutions of the Vlasov-Maxwell

system for an asymmetric current sheet. I would like to adapt this solution to be consistent with periodic boundary conditions of the ViDA code.
4. improving the initial condition of the 2D magnetic reconnection simulations by implementing the spatial variation of the temperature across the current sheet, which is uniform at present. This modification is especially needed in the case of simulations modeling asymmetric magnetic reconnection taking place at the magnetopause.

In general, I would like to use the ViDA code to perform numerical simulations that could closely mimic the observational events at the magnetopause and magnetotail in terms of e.g. current sheet width, temperature and density variation across the boundary, guide field. Indeed, both high resolution observations and recent numerical simulations are revealing the complexity of the magnetic reconnection process at the electron scale and only with a close synergy between numerical simulations and in situ observations we could progress in the understanding of such a complex process.

References

O. Allanson, F. Wilson, T. Neukirch, Y.-H. Liu, J.D.B. Hodgson, Geophys. Res. Lett. **44**, 8685 (2017)
J. Birn et al., J. Geophys. Res. **106**, 3715 (2001)
J. Jara-Almonte, W. Daughton, H. Ji, Phys. Plasmas **21**, 032114 (2014)

Appendix A
Coordinate Systems

In order to facilitate the reading, we report here the coordinate systems used in this Thesis and we provide some terminology linked to the magnetosphere .

The terms *northern* and *southern* are relative to the Earth's equator. Defining the x axis as directed from the Earth to the Sun, the term *noon* indicates the magnetosphere facing the Sun while *midnight* indicates the region beyond the Earth moving along $-x$ direction (Fig. A.1a). The terms *dawn* and *dusk* correspond to 6 a.m. and 6 p.m., respectively. Earth rotates counterclockwise when looking down from the north pole, so when looking from the Sun to Earth, dawn is on the left and dusk is on the right.

Geocentric Solar Ecliptic system (GSE): The origin is at the center of Earth. The x axis points from the Earth towards the Sun. The y axis and the x axis lie in the ecliptic plane and the y axis points towards the dusk. The z axis is perpendicular to the ecliptic plane and directed towards the Northern ecliptic pole.

Geocentric Solar Magnetospheric system (GSM): The origin is at the center of Earth. The x axis points from the Earth towards the Sun. The xz plane contains the Earth's dipole axis, the μ axis in Fig. A.1a. The y axis is perpendicular to Earth's magnetic dipole and it points from dawn to dusk.

De-spun Body Spacecraft Coordinate System (DBCS): This coordinate system is defined by the spacecraft. It is very close to Geocentric Solar Ecliptic (GSE) coordinates (within 2°–3°) since the MMS spacecraft maintains an equatorial orbit.

Local Boundary Normal Coordinate System: This coordinate system is defined locally at the magnetopause or at any other magnetic (or current) structure and it provides the local orientation of the layer. The three orthogonal axis are indicated with the letters L, M, N. In the context of a reconnecting current sheet , the N direction is directed normal to the current sheet while the L and M are tangential to the layer, as it can be seen for the magnetopause current sheet in Fig. A.1b.

G. Cozzani, *Microphysics of Magnetic Reconnection in Near-Earth Space*, Springer Theses, https://doi.org/10.1007/978-3-030-56142-0

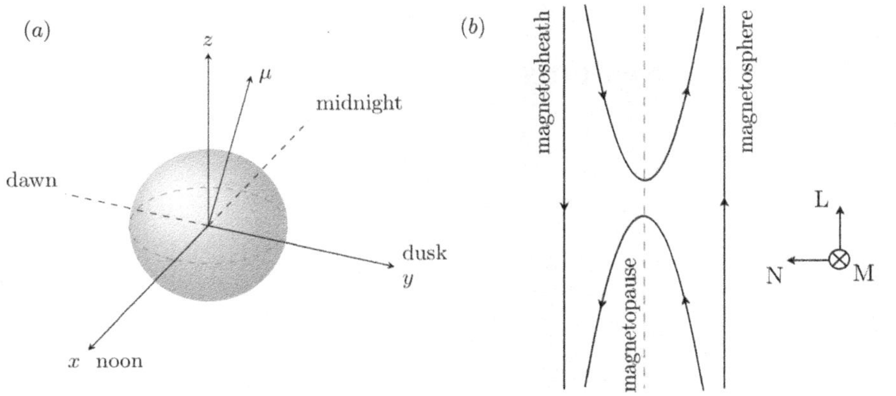

Fig. A.1 **a** Schematic representation of the Earth in the GSE coordinate system. μ indicates the Earth's dipole. **b** Schematic representation of the reconnecting magnetopause current sheet in the boundary local coordinate system LMN

Appendix B
Definition of Electron Agyrotropy

The pressure tensor $\overline{\mathbf{P}}_\alpha$ of the species α is defined as s

$$\overline{\mathbf{P}}_\alpha = m_\alpha \int (\mathbf{v}_\alpha - \langle \mathbf{v}_\alpha \rangle)(\mathbf{v}_\alpha - \langle \mathbf{v}_\alpha \rangle) f(\mathbf{x}_\alpha, \mathbf{v}_\alpha) d\mathbf{v}_\alpha \tag{B.1}$$

where $f(\mathbf{x}_\alpha, \mathbf{v}_\alpha)$ is the distribution function, m_α is the mass and $\langle \mathbf{v}_\alpha \rangle = \frac{\int \mathbf{v}_\alpha f(\mathbf{x}_\alpha, \mathbf{v}_\alpha) d\mathbf{v}_\alpha}{\int f(\mathbf{x}_\alpha, \mathbf{v}_\alpha) d\mathbf{v}_\alpha}$ is the mean velocity. From now, the subscript α is dropped for brevity. In general, the pressure tensor has six independent elements

$$\overline{\mathbf{P}} = \begin{pmatrix} P_{11} & P_{12} & P_{13} \\ P_{21} & P_{22} & P_{23} \\ P_{31} & P_{32} & P_{33} \end{pmatrix} . \tag{B.2}$$

The presence of a magnetic field imposes a preferred direction and it can produce a cylindrically symmetric, or gyrotropic, distribution which corresponds to a pressure tensor of the type

$$\overline{\mathbf{P}} = \begin{pmatrix} P_{||} & 0 & 0 \\ 0 & P_\perp & 0 \\ 0 & 0 & P_\perp \end{pmatrix} . \tag{B.3}$$

Departures from gyrotropy are indicative of the fact that out-of-diagonal elements of the pressure tensor become important. This is crucial in the frame of magnetic reconnection because it indicates that the Ohm's law term which contains the divergence of the pressure tensor may sustain the electric field (see Eq. (2.18)), meaning that electron kinetic processes are playing a role. For this reason, it is useful to find a quantity, and possibly a scalar, that quantifies departures from gyrotropy.

G. Cozzani, *Microphysics of Magnetic Reconnection in Near-Earth Space*, Springer Theses, https://doi.org/10.1007/978-3-030-56142-0

An example is provided by Swisdak (2016) which defines agyrotropy $\sqrt{Q}$ as

$$\sqrt{Q} = \frac{P_{12}^2 + P_{13}^2 + P_{23}^2}{P_{\perp}^2 + P_{\perp} P_{||}} \tag{B.4}$$

where P is the electron pressure tensor and $P_{\perp}$ and $P_{||}$ are the diagonal terms of the tensor written using a set formed by a magnetic field-aligned unit vector and two magnetic field-orthogonal vectors. For gyrotropic tensors $\sqrt{Q} = 0$, while for maximal departures from gyrotropy $\sqrt{Q} = 1$. Other quantities with the same purpose have been proposed by Aunai et al. (2013) and Scudder and Daughton (2008). In particular, Scudder and Daughton (2008) consider a pressure tensor of the type

$$\overline{\mathbf{P}} = \begin{pmatrix} P_{||} & 0 & 0 \\ 0 & P_{\perp,1} & 0 \\ 0 & 0 & P_{\perp,2} \end{pmatrix} \tag{B.5}$$

and define the agyrotropy $A\emptyset$ as

$$A\emptyset = \frac{2|P_{\perp,1} - P_{\perp,2}|}{P_{\perp,1} + P_{\perp,2}}. \tag{B.6}$$

The maximum and minimum value of $A\emptyset$ are 0 (gyrotropy) and 2 (maximum departure from gyrotropy).

Another measure of agyrotropy, called D_{ng} was proposed by Aunai et al. (2013) starting from a tensor written in the same way as Swisdak (2016)

$$D_{ng} = \frac{\sqrt{8(P_{12}^2 + P_{13}^2 + P_{23}^2)}}{2P_{\perp}^2 + P_{||}}. \tag{B.7}$$

It has been shown that $\sqrt{Q}$ definition includes the largest variety of possible agyrotropy sources (Swisdak 2016). For this reason, $\sqrt{Q}$ is adopted in order to quantify agyrotropy throughout this Thesis.

It is worth to underline that these parameters should be used with care. Indeed, out-of-diagonal elements of the pressure tensor usually increase when the plasma density is as low as the typical magnetospheric values and when the electron temperature is anisotropic with $T_{e,||} > T_{e,\perp}$. This conditions are commonly found in the outer magnetosphere . This behavior of $\sqrt{Q}$ has been reported by Argall et al. (2018).

References

N. Aunai, M. Hesse, M. Kuznetsova, Phys. Plasmas **20**, 092903 (2013)
M.R. Argall et al., J. Geophys. Res. **123**, 146 (2018)
J.D. Scudder, W. Daughton, J. Geophys. Res. **113**, A06222 (2008)
M. Swisdak, Geophys. Res. Lett. **43**, 43 (2016)

Bibliography

W. Gonzalez, E. Parker (eds.), *Magnetic Reconnection. Concepts and Applications* (Springer, Cham, 2016)

Index

G. Cozzani, *Microphysics of Magnetic Reconnection in Near-Earth Space*, Springer Theses, https://doi.org/10.1007/978-3-030-56142-0

S

T

W

X

The manufacturer's authorised representative in the EU is Springer Nature Customer Service Centre GmbH, Europaplatz 3, 69115 Heidelberg, Germany. If you have any concerns regarding our products, please contact ProductSafety@springernature.com

Printed and bound by CPI Group (UK) Ltd, Croydon, CR0 4YY
30/09/2026
02225515-0017